Understanding Lasers

Jeff Hecht

Contributing Editor
Laser & Optronics

HOWARD W. SAMS & COMPANY

A Division of Macmillan, Inc.
4300 West 62nd Street
Indianapolis, Indiana 46268 USA

International Standard Book Number: 0-672-27274-1
Library of Congress Catalog Card Number: 88-61547

Acquisitions Editor: *Greg Michael*
Development Editor: *Jennifer L. Ackley*
Illustrator: *Don Clemons*
Cover
  Concept: *DGS&D Advertising*
  Photography: *Cassell Productions, Inc.*
  Components: *Melles Griot, Irvine, CA*
Composition: *Shepard Poorman Communications Corp.*

Printed in the United States of America

Contents

Preface

I suspect more nonsense has been written about lasers than about any of the other "space age" technologies of the last half of the twentieth century. Even the technologically unsophisticated have a general idea of what electronics and spacecraft can do. Ask them about lasers, however, and they start spouting misconceptions about death rays and ray guns—stuff of science-fiction stories. You won't hear about them here. What you will learn about is the exciting world of real lasers—how they work, and what they can do.

Two decades ago, one irreverant observer called the laser, "a solution looking for a problem." There was truth to that assessment, because scientists learned how to make lasers work before engineers learned how to put them to work. Now that has changed. Lasers are a part of our lives in more ways then we realize. Listen to a compact disc audio recording, and you hear crystal-clear music read by a laser beam. Make a long-distance phone call, and chances are excellent that your voice travels part of the way via laser beam. Go to the supermarket and a laser beam will read the prices at the checkout counter. Elsewhere, lasers are doing hundreds of other jobs, in places ranging from hospitals and research laboratories to factories and construction sites.

This book will introduce you to real-world lasers and how they work. It is arranged somewhat like a textbook. Each chapter starts by saying what it will cover, ends by reviewing key points, and is followed by a short multiple-choice quiz. It builds understanding step-by-step. We start with a broad overview of lasers. In the second chapter, we review key concepts of physics and optics that are essential for you to understand lasers. You should skim this even if you have a background in physics, especially to check basic optical concepts. The third and fourth chapters cover key concepts of how lasers work. In the fifth chapter, we take a quick look at all the accessories that help lasers do their jobs better. Try to master each of these chapters before going on to the next and those that follow.

The sixth through ninth chapters go into more detail on the workings of specific types of lasers. We've given special attention to the hottest of the commercial technologies, semiconductor lasers, in Chapter 8. Two of the most exciting lasers still in the laboratory—free-electron and X-ray lasers—are covered in Chapter 9. Chapters 10 and 11 give a brief overview of laser applications. The glossary and index are included to help make this book a useful reference after you have studied it.

To keep this book a reasonable length, we concentrate on lasers and their workings. We cover optics and laser applications only in brief, but after reading this book you may want to study them in more detail. If you are interested in fiber-optic communications, you may want to read another book I wrote, *Understanding Fiber Optics*.

I met my first laser two decades ago, and have been writing about laser technology since 1974. I have found it fascinating, and I hope you will, too.

JEFF HECHT

Acknowledgment

Thanks to Dan Hewak for comments on early drafts of some chapters.

Introduction & Overview

ABOUT THIS CHAPTER

This chapter will introduce you to lasers. It will give you a basic idea of how they work, how they are used, and what properties are important. This basic understanding will serve as a foundation for the more detailed descriptions of lasers in later chapters.

LASERS IN FACT AND FICTION

The laser sits near the top of any list of the greatest inventions of the last half of the twentieth century. Together with the satellite, the computer, and the integrated circuit, it is a symbol of "high technology." Like the other technologies, lasers affect our lives in many ways, and are growing steadily in importance. Laser technology is both fascinating in itself and an important tool in fields from medicine to communications. However, few people outside the field understand it well.

One tale summarizes the problem well. In 1962, a now-defunct supplement to Sunday newspapers published an article on "The Incredible Laser." Its wild predictions of laser cannons being "just around the corner" amused laser physicists, who knew better. One of them, Arthur L. Schawlow—who later shared in the 1981 Nobel Prize in Physics for his research with lasers—posted the article on his door at Stanford University, along with the words, "for credible lasers, see inside."

Over the years, Schawlow has given away many copies of the poster, and it has amused many students and laser engineers and scientists. Yet the basic point remains true. Too many people think lasers are science-fictional ray guns or death rays. Military researchers are trying to develop laser weapons, but so far they only have laboratory prototypes. Even in the optimistic plans of the Strategic Defense Initiative, lasers are years away from deployment.

1

Real laser technology may not be as spectacular, but it is important and fascinating. Lasers can send signals through miles of fiber-optic cable, print computer output, read printed codes in the supermarket, diagnose and cure disease, cut and weld materials, and make ultraprecise measurements. With laser light you can record three-dimensional holograms, project ever-changing light patterns, spot flaws in a centuries-old painting, or play crystal-clear digital music recorded on a compact disc. This book will teach you how this real laser technology works.

WHAT IS A LASER

The word laser was coined as an acronym, for *L*ight *A*mplification by the *S*timulated *E*mission of *R*adiation. The word tells us that laser light is special light. Ordinary light, from the sun or a light bulb, is emitted spontaneously, when atoms or molecules get rid of excess energy by themselves, without any outside intervention. Stimulated emission is different, because it occurs when an atom or molecule holds onto excess energy until it is "stimulated" to emit it as light. We'll come back to look more closely at stimulated emission in Chapter 3.

Albert Einstein was the first to suggest the existence of stimulated emission in a paper published in 1917. However, for many years physicists thought that atoms and molecules always were much more likely to emit light spontaneously, and that stimulated emission thus always would be much weaker. It was not until after World War II that physicists began trying to make stimulated emission dominate. They sought ways that one atom or molecule could stimulate many others to emit light, amplifying it to much higher powers.

The first to succeed was Charles H. Townes, then at Columbia University. Instead of working with light, however, he worked with microwaves which have a much longer wavelength. He built a device he called a "maser," for *M*icrowave *A*mplification by the *S*timulated *E*mission of *R*adiation. He thought of the key idea in 1951, but the first maser was not completed until a couple of years later. Before long, many other physicists were building masers, and trying to find how to produce stimulated emission at even shorter wavelengths.

The key concepts emerged about 1957. Townes and Arthur Schawlow, then at Bell Telephone Laboratories, wrote a long paper outlining the conditions needed to amplify stimulated emission of visible light waves. At about the same time, similar ideas crystallized in the mind of Gordon Gould, then a 37-year-old graduate student at Columbia, who wrote them down in a series of notebooks. Townes and Schawlow published their ideas in a scientific journal, *Physical Review Letters*, but Gould filed a patent application.

Three decades later, people still argue who deserves credit for the concept of the laser. Townes and two Soviet maser pioneers, Nikolai Basov and Aleksander Prokhorov, shared the 1964 Nobel Prize in physics for their pioneering work on "the maser/laser principle." Schawlow eventually shared the 1981 Nobel Physics Prize for research done with lasers. For many years, Gould had to settle only for credit for coining the word "laser" in his notebooks. However, after nearly two decades of legal struggles and comparative obscurity, he finally received three patents worth millions of dollars on some important aspects of laser technology. The last of those patents was issued in November 1987, 30 years after Gould sat down to write his notebooks.

Publication of Townes and Schawlow's paper stimulated widespread efforts to make lasers. Schawlow and others went to work at Bell Laboratories. Gould took his ideas to a small Long Island company, TRG Inc., which used them in a research proposal to the Department of Defense. The Pentagon brass was enthusiastic about the idea, but not about Gould's brief involvement with a Marxist study group a decade earlier, so they wouldn't grant Gould the security clearance he needed to work on the project. Scientists at many other government, industry, and university laboratories also tried to make lasers.

The winner of the great laser race, on May 16, 1960, was a dark horse, Theodore Maiman. He had decided that synthetic ruby was a good laser material, even though Schawlow had publicly said it wouldn't work. Maiman's managers at Hughes Research Laboratories in Malibu, California had told him to stop wasting his time trying to make a ruby laser. But Maiman continued, and by putting mirrors on each end of a ruby rod and illuminating it with a bright flashlamp, he made the world's first laser, shown in *Figure 1-1*. (You may see some other pictures of the "first laser" in which the laser looks larger. Those are courtesy of a Hughes Research Labs photographer, who thought the first laser was too small and had Maiman pose with a larger version that he operated later.)

Figure 1-1. Theodore Maiman holds the world's first laser.
(Courtesy Theodore Maiman)

Many other types of laser followed, but Maiman's ruby laser was typical in many ways of what we expect a laser to be. It emitted light in a narrow, tightly concentrated beam. The light was all the same wavelength—694 nanometers at the long-wavelength end of the red spectrum. The light waves were coherent; that is, they all aligned with each other and marched along in step.

Maiman's laser emitted light in short, intense pulses; many other lasers also operate in pulsed mode. However, other lasers emit steady beams. Power levels can span a wide range. Some lasers emit less than a milliwatt (0.001 watt) of light; others emit steady beams of many kilowatts. The highest laser powers—trillions of watts—are achieved in ultrashort pulses lasting only about a billionth of a second. *Figure 1-2* shows a few representative lasers.

(A) Helium-neon laser. *(Courtesy Melles Griot)*

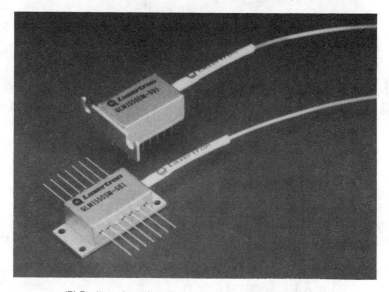

(B) Packaged semiconductor laser. *(Courtesy Lasertron Inc.)*

Figure 1-2. Representative lasers.

(C) Solid-state laser. *(Courtesy Lasermetrics Inc.)*

(D) Carbon-dioxide laser. *(Courtesy Coherent Inc.)*

Figure 1-2. Cont.

We think of lasers as emitting visible light, but the basic principles behind the laser are valid through much of the electromagnetic spectrum. We already saw that Townes produced the first stimulated emission in the microwave spectrum. Many lasers emit infrared and ultraviolet light, and researchers now are pushing lasers into the X-ray spectrum.

Maiman's first laser was small enough for him to hold in one hand, but the size of lasers can vary widely. Military laboratories have built monstrous lasers that fill the interior of good-sized buildings. Yet the most common type of laser is made of semiconductor material, and is only the size of a grain of salt.

We'll learn more about the properties of lasers in later chapters. For now, it is most important to remember that the types of lasers are many and varied.

HOW LASERS ARE USED

In the early 1960s, it seemed that every physicist and engineer wanted to get his or her hands on a laser and play with it. They aimed lasers at just about everything they could find. They shot so many holes in razor blades that for a while laser power was informally measured in "gillettes." Yet there were few practical applications, and for a time the laser seemed to be, in the words of one wag, "a solution looking for a problem."

That is no longer the case. Lasers are now used for many purposes in many fields. A sampling of laser applications follows. Chapters 10 and 11 describe some of the uses in more detail.

Reading and Writing Information

Playing audio compact discs

Playing videodisks

Reading Universal Product Code in stores

Reading computer data on optical disks

Laser printers for computer output

Measurement and Inspection

Projecting straight lines for construction alignment and irrigation

Measuring the range to distant objects

Measuring small distances very precisely

Illuminating cells for biomedical measurements

Laser-induced fluorescence measurements

Studies of atomic and molecular physics

Measuring concentrations of chemicals or pollutants

Detecting flaws in aircraft tires

Medicine
Treatment of diabetic retinopathy to forestall blindness
Laser surgery
Laser bleaching of port wine stain birthmarks
Shattering of kidney stones

Materials Working
Cutting, drilling, and welding plastics, metals, and other materials
Cutting cloth
Drilling materials from diamonds to baby-bottle nipples
Engraving wood
Marking identification codes

Military Applications
Antisensor weapons
Antisatellite weapons
Antimissile weapons
Battle simulation
Pinpointing targets for bombs and missiles
Simulating effects of nuclear weapons

Other Applications
Making holograms
Laser light shows
Displays
Laser pointers
Controlling chemical reactions
Changing concentrations of isotopes
Producing nuclear fusion
Basic research

We can get a better feeling for how lasers are used by breaking down major applications into families. One large family of applications uses lasers as highly directional light bulbs for reading, sending signals or making measurements. Another uses the energy in laser light to change things, ranging from writing spots on paper to drilling holes in diamond. Other applications rely on the special properties of laser light, such as the way its light waves are aligned, or the fact that it consists of only a single color.

Low-power lasers may seem to do the same job as a light bulb, but they can do it better. For example, supermarkets use laser scanners under checkout counters to read bar codes on packages. The bar code identifies each product to the store's computer, which recalls the price and prints it out at the terminal. Why use a laser? The red laser light is in a thin, highly directional beam, so it can be focused onto packages at various distances above the counter. The scanner includes a filter that cuts out all light other than the single shade of red emitted by the laser, reducing the likelihood of errors caused by scattered light from overhead lamps. Engineers looked at many other alternatives, and decided the laser was best.

Semiconductor lasers are widely used because they are very small and can produce a few milliwatts of light that can be tightly focused to a very small spot. That is ideal for many uses, ranging from playing music on compact discs to sending signals through optical fibers. What's more, by varying the current passing through the laser, you can control how much light it emits—ideal for communications or writing computer output.

Other lasers produce short, intense pulses of light. If these pulses are focused, they can concentrate a lot of energy on a tiny spot. Such pulses can make tiny marks on a metal surface or drill holes. Because laser pulses don't get dull, they are very good for drilling hard materials, like titanium metal or even diamond. Laser pulses are also very good for drilling into soft material, like rubber or many plastics, because they don't twist, bend or turn the material. A series of pulses, or a higher-power steady beam, can cut or weld. Physicians can carefully control laser energy to remove diseased tissue without damaging healthy cells. Computers can control cutting and drilling lasers in automated machining centers.

The alignment of laser light waves—called "coherence" makes laser light ideal for holography, the recording and reconstruction of three-dimensional images. Holograms make striking artistic displays, projecting three-dimensional images that look so real you want to reach out and touch them. They also serve more practical purposes.

You can measure tiny deformations in an object by recording two holograms on the same piece of film, one before a stress is applied, and one after. Illuminate the film to produce an image, and you can see where the object moved. This lets inspectors spot flaws in aircraft tires before they are put on a plane, so you can ride more safely.

In later chapters, we will delve more into laser applications. We also will see that new applications are being developed steadily. Government and industrial laboratories are developing laser techniques to perform tasks ranging from treatment of cancer to controlling nuclear fusion.

IMPORTANT LASER PROPERTIES

Lasers take many different shapes. Other characteristics, such as output wavelength and power level, also differ among lasers. Different types of lasers also share some common properties, such as the concentration of their output energy in a narrow beam. We list many important properties of lasers below. Each of these deserves a bit more explanation.

Major Laser Properties

- Wavelength(s)
- Output Power
- Duration of emission (pulsed or continuous)
- Beam divergence and size
- Coherence
- Efficiency and power requirements

Wavelength(s)

As we will see later, wavelength is a fundamental characteristic of visible light and other forms of electromagnetic radiation. We sense the wavelength of visible light as color (although there are some complications that make the sensing inexact). Each type of laser emits a characteristic wavelength or range of wavelengths. The wavelengths depend on the type of material that emits the laser light, the laser's optical system, and the way the laser is energized. Laser action can produce infrared, visible, and ultraviolet light; "masers" produce microwaves by essentially the same process. If you want a particular wavelength, your choice usually is limited to no more than a few types of lasers. *Table 1-1* lists a sampling of important laser types and their wavelengths in nanometers (1 nm = 10^{-9} meter)

Table 1-1. Important Laser Types and Their Wavelengths

Type	Wavelength (nm)
Krypton-fluoride excimer	249
Xenon-chloride excimer	308
Nitrogen gas (N_2)	337
Organic dye (in solution)	300–1000 (tunable)
Krypton ion	335–800
Argon ion	450–530 (488 and 514.5 strongest)
Helium neon	543, 632.8, 1150
Semiconductor (GaInP family)	670–680
Ruby	694
Semiconductor (GaAlAs family)	750–900
Neodymium YAG	1064
Semiconductor (InGaAsP family)	1300–1600
Hydrogen-fluoride chemical	2600–3000
Carbon dioxide	9000–11000 (main line 10,600)

Most lasers are "monochromatic," a term that means they emit only one wavelength. Actually, their emission is a range of wavelengths, but the range is so narrow that for most purposes it seems to be a single wavelength. Some of these lasers emit light at different wavelengths under different conditions. For example, the helium-neon laser is best known for its red output at 632.8 nanometers. However, the same gas mixture can be used with different optics to emit green light at 543 nanometers, or to emit infrared light.

Some lasers can emit two or more wavelengths at once; this is called *multiline operation*. In some cases, the wavelengths are close together. *Table 1-1* shows that the argon laser emits several wavelengths between 450 and 530 nm, with the strongest lines at 514.5 and 488 nm. In others, they are widely separated; the krypton laser emits several different-colored visible wavelengths. Some lasers emit many wavelengths in a limited range; one example is the carbon-dioxide laser, with dozens of wavelengths between 9 and 11 micrometers (9000 to 11,000 nm) in the infrared.

As we will see later, the way in which lasers emit light tends to concentrate output energy at specific wavelengths. However, some lasers can emit light over a comparatively wide range of wavelengths. The most important examples are lasers in which light is emitted by organic dyes in solution. They can be tuned in wavelength by adjusting the laser's optical system.

Output Power

Output power measures the strength of a laser beam, which differs widely among lasers. Strictly speaking, power is the flow of light energy from the laser in the form of the laser beam; it is measured in watts and defined by the formula:

$$Power = d(energy) / d(time)$$

Laser output powers cover a wide range. Some lasers produce beams containing less than a thousandth of a watt (a milliwatt). Others produce thousands of watts (kilowatts).

You cannot make one laser produce the whole range of output powers just by turning a knob. Some lasers can be adjusted over a limited power range, but others are designed to emit a stable power level. Some types of lasers cannot be scaled to high power levels. For example the helium-neon laser is not able to produce more than a few dozen milliwatts. Only a few types, notably carbon-dioxide and chemical lasers, can produce thousands of watts in a steady beam. If you need high laser powers, you often don't have much choice among types of lasers—and thus your choice of wavelengths is limited.

Pulsed and Continuous Output

You may think of lasers as emitting a steady beam of light, like a light bulb. Many, including the red helium-neon laser you are most likely to have encountered in a laser demonstration (or in a supermarket or construction site), do. However, others emit pulses of light.

Pulses come in various durations and repetition rates. The length or duration of a pulse can range from milliseconds to femtoseconds—or, in scientific notation, from 10^{-3} to 10^{-15} second. The pulses may be repeated once a minute, or may appear thousands or even millions of times in a second. (Because the eye's response is much slower than the laser's, more than a few dozen pulses per second look continuous to the eye but electronic detectors can recognize those short pulses.)

There are important and fundamental relationships among pulse length, energy, repetition rate, and power. A laser may have extremely high peak power during a short pulse but, because the pulse is short, it doesn't contain much energy. To make a simple approximation:

$$\text{Pulse Energy} = \text{Peak Power} \times \text{Pulse Length}$$

If you wanted to make the relationship exact, you would have to take an integral:

$$\text{Energy} = \int \text{Power} \, d(\text{Time})$$

because power varies with time during the pulse. However, simple multiplication usually will give a reasonable estimate.

The average power in a pulsed laser beam is different from the peak power; it is a measure of the average energy flow per second. Thus it equals:

$$\text{Average Power} = \frac{\text{Number of Pulses} \times \text{Pulse Energy}}{\text{Time}}$$

If the number of pulses is measured as repetition rate, time is normalized to one second, and the average power becomes:

$$\text{Average Power} = \text{Repetition Rate} \times \text{Pulse Energy}$$

Peak and average power both can be important quantities, depending on the laser application. Both are measured in watts, but as would be expected, peak power is higher. Pulse energy is measured in joules. The conversion between the two units is

$$Watts = Joules/Seconds$$

or

$$Joules = Watts \times Seconds$$

Beam Divergence and Size

If you see a laser beam shining through dusty air, it looks as thin as a string or pencil line. However, if you look carefully at how the beam travels, you will see that it has a certain diameter, and that the farther the beam goes, the larger it gets. You can see the same effect more readily in the beam from a flashlight or searchlight. This spreading is called *divergence*.

Beam divergence typically is measured in milliradians, or thousandth of a radian. A radian equals 57.3 degrees; 2π radians equals 360 degrees, or a full circle. Radians are used because the size of a small angle in radians is almost equal to its sine, making it simple to calculate the size of a laser beam. Once you get far enough from the laser, you just multiply the sine of the divergence angle (which equals the value of the angle for small angles measured in radians) by the distance the beam has travelled to get the spot radius, as shown in *Figure 1-3*. Thus,

$$Radius = Distance \times \sin(Beam\ Divergence)$$

If the divergence is 2 milliradians, after the beam has travelled 10 meters, its radius is 20 mm, and its diameter is 40 mm. After it has travelled 100 feet, its diameter is 0.4 feet. Not all lasers have such small divergence, but those that do can send their beams long distances. This high directionality of laser beams is one of their most important properties.

Coherence

Light waves are said to be *coherent* if they are all in phase with one another. *Figure 1-4* compares coherent and incoherent light waves. The peaks and valleys of the coherent light waves in *Figure 1-4A*

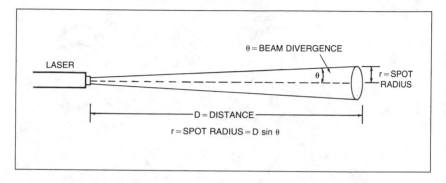

Figure 1-3. Calculating the size of a laser spot from the beam divergence.

are all lined up with each other. The peaks and valleys of incoherent light waves (*Figure 1-4B*) do not line up. Laser light is coherent; ordinary light from the sun, an incandescent bulb, or a fluorescent tube, is incoherent.

Take a close look at *Figure 1-4*, and you will note something else about coherent light. If it is to remain coherent over a long distance, all the light waves must have the same wavelength. Suppose one light wave is just 0.1% longer than another; after 500 wavelengths, the peaks of one wave will line up with the valleys of the other, so the light waves will be out of phase and incoherent. Thus the more monochromatic a laser is, the more coherent it is. Because the wavelengths of visible light are very small (500 nm or 5×10^{-7} meter), light must be quite monochromatic to be coherent over a long distance. If visible light is coherent over 500 wavelengths, the distance is only 2.5×10^{-4} m—a quarter of a millimeter. Monochromatic light need not be coherent, but light that is not monochromatic cannot stay coherent over a long distance.

Many laser applications do not require coherent light; the coherence is incidental. Coherent light is essential, however, for holography and some measuring applications. The reasons for this will be described later.

Efficiency and Power Requirements

A laser's wavelength, output power, coherence, and pulse characteristics all are important in picking a laser. Two other factors also enter the picture—efficiency and power requirements. Lasers

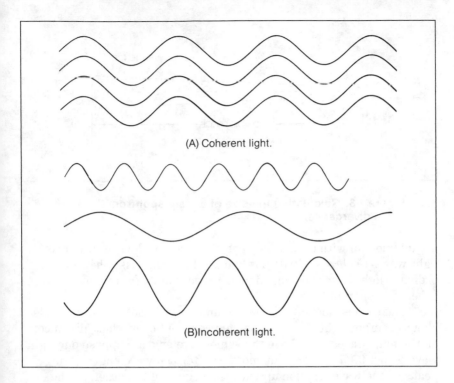

(A) Coherent light.

(B)Incoherent light.

Figure 1-4. Coherent and incoherent light.

differ widely in how efficiently they convert input energy (usually in the form of electricity) into light. Like other light sources, they are not very efficient in generating light, with the best converting up to about 20% of input energy into light. Many types convert as little as 0.01% or even 0.001% of the input energy into light.

Efficiency becomes a more important consideration at higher output powers. It's not a big problem if a 1-milliwatt laser produces a watt of waste heat, because it's easy to dissipate. However, it would be very hard to dissipate the million watts of waste heat produced if a one-kilowatt (1000-watt) laser operated at the same efficiency.

LASERS, PHYSICS, AND OPTICS

The internal operation of lasers depends on physics. Laser emission depends on how atoms and molecules emit light, and that, in turn, depends upon their internal structure. To understand the internal structure of lasers, and how it affects the ways in which they emit light, you will need to take the brief introductory tour of modern physics in Chapter 2.

The way lasers work also depends intimately on optics. We will also cover some laws of optics in Chapter 2, and touch on optics repeatedly throughout the book. There is no escaping from optics. Lasers are optical devices, and to use lasers you need optics, ranging from the mirrors within the laser itself, to lenses and mirrors that focus and direct laser beams.

1

What Have We Learned

- Laser light is produced by stimulated emission.
- The microwave-emitting maser preceded the laser.
- 30 years after the laser was invented, people still argue who deserves credit.
- Theodore Maiman made the first laser by exciting a ruby rod with a flashlamp.
- The laser was once called "a solution looking for a problem," but it has since found many applications.
- Low-power lasers are good light sources for reading and writing.
- Short, intense laser pulses can mark surfaces or drill holes.
- Different types of lasers share many properties.
- Wavelength is a fundamental characteristic of all light.
- Laser light is usually monochromatic.
- Power is the flow of light energy.
- Lasers emit different power levels.
- Lasers can emit pulsed or continuous light.
- Average power equals pulse repetition rate times pulse energy.
- A laser beam has a characteristic spreading angle (divergence) and diameter.
- Most lasers emit "coherent" light waves that are in phase with each other and have the same wavelength.
- Efficiency and power requirements are important in picking lasers.
- The internal operation of lasers depends on basic laws of physics.

WHAT'S NEXT

In Chapter 2, we will introduce the basic physical concepts behind lasers. Much of the material is basic, and parts may seem repetitive if you have been exposed to physics before. However, you should at least scan through quickly to make sure you understand the key concepts. They will reappear repeatedly in the following chapters.

Quiz for Chapter 1

1. The word laser originated as:
   a. A military codeword for a top-secret project
   b. A trademark
   c. An acronym for Light Amplification by the Stimulated Emission of Radiation
   d. The German word for light emitter

2. The first laser was made by:
   a. Charles Townes
   b. Theodore Maiman
   c. Gordon Gould
   d. Nikolai Basov and Aleksander Prokhorov
   e. H. G. Wells

3. The first laser emitted
   a. Pulses of 694-nm red light
   b. A continuous red beam
   c. Pulses of white light from a helical flashlamp
   d. Spontaneous emission

4. Laser light is which of the following:
   a. Coherent
   b. Stimulated emission
   c. Spontaneous emission
   d. Monochromatic
   e. a, b, and d
   f. c and d

5. Which of the following is NOT a laser application
   a. Printing computer output
   b. Stimulating rainfall from clouds
   c. Making very precise measurements
   d. Playing compact disc audio recordings
   e. Engraving wood

6. The coherence of laser light is important for:
   a. No practical applications
   b. Drilling holes
   c. Getting laser light to pass through air
   d. Holography
   e. None of the above

7. Which important laser emits light in the visible range, 400 to 700 nm?
   a. Argon ion
   b. Nitrogen
   c. Carbon dioxide
   d. Neodymium YAG
   e. Chemical

8. Which is the proper measurement of average power emitted by a pulsed laser?
   a. Energy × Time
   b. Pulse Energy × Repetition Rate
   c. Pulse Energy / Repetition Rate
   d. Peak Power × Pulse Length
   e. None of the above

9. How do you calculate the radius of a laser spot at a given distance if you know beam divergence?
   a. Multiply beam divergence in degrees by distance in milliradians
   b. Divide beam divergence in degrees by distance in meters
   c. Measure it with a ruler
   d. Multiply the sine of the beam divergence by distance in meters
   e. Multiply power in watts by beam divergence

10. What type of light can be coherent?
    a. Spontaneous emission
    b. Monochromatic and in phase
    c. Narrow beam divergence
    d. Monochromatic only

Physical Basics

ABOUT THIS CHAPTER

Lasers are not isolated. They are one of the many fruits of the revolution that opened up new realms of physics in the early twentieth century. To understand lasers, you need a basic understanding of light, optics, quantum mechanics, and some other aspects of physics. This chapter will give you the basic background in modern physics you will need to understand how lasers work.

ELECTROMAGNETIC WAVES AND PHOTONS

Early physicists had long and loud debates over the nature of light. Isaac Newton held that light was made up of tiny particles. Christian Huygens believed that light was made up of waves, vibrating up and down in a direction perpendicular to the direction the light travels. Newton's theory came first, but Huygens' theory explained early experiments better, so for a long time it was assumed to be right.

Today, we know that both theories are right—in part. Much of the time, light behaves like a wave. It is called an electromagnetic wave, because it is composed of both electric and magnetic fields. As shown in *Figure 2-1*, both those fields oscillate perpendicular to the direction the light wave is travelling, and perpendicular to each other. Light waves are called transverse waves, because they oscillate in a direction transverse to the direction they travel.

Certain characteristics identify waves. One is wavelength, the distance between successive peaks, illustrated in *Figure 2-1*; it is measured in units of length. Another is frequency, the number of wave peaks passing a point in a second. Frequency is measured in cycles (peaks) per second, a quantity called hertz (named after Heinrich Hertz, the 19th-century discoverer of radio waves). Multiply

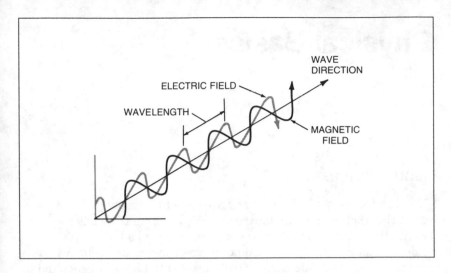

Figure 2-1. Structure of an electromagetic wave.

the two units together, and you get the wave's speed, in units of length per second.

$$\text{Wavelength} \times \text{Frequency} = \text{Speed}$$

The speed of light in a vacuum is a universal constant, 300,000 kilometers per second, or 186,000 miles per second, so this product is always a constant for light and other electromagnetic waves. This means that the wavelength is inversely proportional to frequency— the shorter the wavelength, the higher the frequency. If we use the standard symbols for wavelength (λ), frequency (ν) and the speed of light (c), we can write simple equations to relate wavelength and frequency:

$$\lambda = c \,/\, \nu$$
$$\nu = c \,/\, \lambda$$

At other times, light acts as if it is made up of particles called photons. A photon is a chunk or quantum of energy. Although Einstein's famous $E = mc^2$ equation tells us that mass and energy are equivalent, photons are considered to be massless particles, and only massless particles can travel at the speed of light.

Each photon has an amount of energy proportional to the

frequency of the light wave, and thus inversely proportional to the wavelength. Photon energy is related to frequency by the formula

$$\text{Energy} = \text{Planck's Constant} \times \text{Frequency}$$

or

$$E = h\nu = 6.63 \times 10^{-34}\nu$$

where,
   h is a constant known as Planck's constant after German physicist Max Planck who discovered the relationship.

Its value is 6.63×10^{-34} joule-second. Combine this formula with the relationship between wavelength and frequency, and you have another expression for wavelength:

$$\text{Wavelength} \times \text{Energy} = \text{Planck's Constant} \times \text{Speed of Light}$$

or

$$\lambda \times E = hc = 1.99 \times 10^{-25} \text{ joule-meter}$$

   As you work with light and other electromagnetic waves, you will come to remember these conversions. For now, the most important lesson is that the three descriptions of light waves are equivalent. Each light wave can be described by a photon energy, a wavelength, or a frequency. The numbers are different, but the descriptions are equivalent. For example, light with a 1-micrometer wavelength has a frequency of 3×10^{14} hertz. Its photon energy is 2×10^{-19} joule. For convenience, physicists often measure photon energy in units called electronvolts, the energy an electron acquires in moving through a potential of one volt. One electronvolt equals 1.6022×10^{-19} joule, so a photon energy of 2×10^{-19} joule equals 1.24 electronvolts (eV), a more convenient unit of measurement.
   It is easy to make mistakes in converting units, so it helps to remember these simple rules of thumb:

- The higher the frequency, the shorter the wavelength
- The higher the frequency, the larger the photon energy
- The shorter the wavelength, the larger the photon energy

Because electromagnetic waves (or photons) are emitted or radiated by various objects, physicists sometimes call them *electromagnetic radiation*. If the word "radiation" sounds distressingly like something that comes from a leaky nuclear reactor, blame it on the way headline writers tend to throw away adjectives. The word radiation covers a very broad range of phenomena, only a few of which are dangerous. Most, like light, infrared radiation, and radio waves, are innocuous parts of the electromagnetic spectrum.

The Electromagnetic Spectrum

We usually think of the spectrum as the rainbow of colors spread out when we pass sunlight through a prism. Those colors are how our eyes see different wavelengths; the sensation of red means longer wavelengths than orange or yellow, for example. Our eyes can see only a narrow range of wavelengths, but the whole electromagnetic spectrum covers a tremendous range, from extremely low frequency waves many miles long to gamma rays a trillionth of a meter long. This range is tabulated in *Table 2-1*:

Table 2-1. Wavelengths and Frequencies of Electromagnetic Radiation

Name	Wavelength (m)	Frequency (Hz)
Gamma Rays*	under 3×10^{-11}	over 10^{20}
X Rays*	3×10^{-11} to 10^{-8}	3×10^{16} to 10^{20}
Ultraviolet light*	10^{-8} to 4×10^{-7}	7.5×10^{14} to 3×10^{16}
Visible light	4×10^{-7} to 7×10^{-7}	4.2×10^{14} to 7.5×10^{14}
Infrared light	7×10^{-7} to 10^{-3}	3×10^{11} to 4.2×10^{14}
Microwaves	10^{-3} to 0.3	10^9 to 3×10^{11}
Radio waves	0.3 to 30,000	10^4 to 10^9
Low-frequency waves*	over 30,000	under 10,000

*No standard boundaries.

In *Table 2-1*, all frequency and wavelength values are expressed in the same (metric) units to simplify comparison. However, in practice the values of frequency, wavelength, and other quantities such as time and power, are expressed in metric units with the standard prefixes shown in *Table 2-2*. You probably already know some of these prefixes, but you are likely to discover others in exploring the world of lasers. Virtually everything optical is measured in metric units, and we will follow that practice in this book. Other books often give the wavelength of light in Angstrom units (Å), which equal 10^{-10} meter or 0.1 nanometer. We will follow the now-standard system of metric units that does not include the Angstrom. If you encounter Angstroms elsewhere, just divide by 10 to get the measurement in nanometers.

The division of the electromagnetic spectrum into various

Table 2-2. Prefixes Used for Metric Units

Prefix	Abbreviation	Meaning	Number
exa	E	quintillion	10^{18}
peta	P	quadrillion	10^{15}
tera	T	trillion	10^{12}
giga	G	billion	10^{9}
mega	M	million	10^{6}
kilo	k	thousand	1000
deci	d	tenth	0.1
centi	c	hundredth	0.01
milli	m	thousandth	0.001
micro	μ	millionth	10^{-6}
nano	n	billionth	10^{-9}
pico	p	trillionth	10^{-12}
femto	f	quadrillionth	10^{-15}
atto	a	quintillionth	10^{-18}

domains is somewhat arbitrary, reflecting the historical accident of the piecemeal discovery of different parts of the spectrum. Early physicists discovered infrared and ultraviolet light by using measurement instruments to look beyond the part of the spectrum visible to the human eye. However, radio waves were discovered independently by Heinrich Hertz in the 19th century. X rays and gamma rays also were discovered independently. Eventually, physicists realized that all types of waves were part of the same family and put them all together. However, some fuzziness remains at the borderlines, especially at those where little research has been done, such as the border between X rays and the ultraviolet.

Even the one place where boundaries might seem well-defined—the limits of visibility to the human eye—is not rigidly defined. The eye's sensitivity to light drops off gradually with wavelength, especially in the infrared. The limits of visibility usually are defined as 400 to 700 nanometers (or, equivalently, 4000 to 7000 Angstroms), but the eye can faintly sense somewhat longer and shorter wavelengths. Thus some books define the visible range as extending from 380 nm to 750 or 760 nm.

Strictly speaking, you might expect the term "light" to apply only to the part of the electromagnetic spectrum sensed by the human eye. However, in practice the definition is broader, and includes much of the infrared and ultraviolet. The reason for the broader definition is that, except for their invisibility to the human eye, the near ultraviolet and near infrared parts of the spectrum behave much like visible light. The same types of optics used for visible light also can be used in the adjacent parts of the infrared and ultraviolet. Because of this common behavior, this part of the spectrum sometimes is called the "optical" region. The properties of electromagnetic waves change with wavelength, and the farther the wavelength is from the visible, the more differently it behaves. For example, air strongly absorbs ultraviolet wavelengths shorter than about 200 nanometers. Wavelengths at the long-wavelength end of the infrared spectrum (often called "submillimeter" waves because their wavelengths are slightly under a millimeter) are reflected by wire-mesh screens, can travel through waveguides, and otherwise behave more like microwaves than visible light. Indeed, if you look at where they are in the spectrum, they should behave more like microwaves. Submillimeter waves are a factor of 10 or 100 shorter than microwaves, but a factor of 1000 longer than visible light.

Lasers normally operate at infrared, visible, or ultraviolet wavelengths. Similar devices operating at microwave wavelengths are called "masers," for Microwave Amplification by the Stimulated Emission of Radiation. As we saw earlier, Charles Townes invented the maser before anyone had a clear idea how to build a laser. Recently, scientists at the Lawrence Livermore National Laboratory have pushed the record for shortest laser wavelength to about 10 nanometers (10^{-8} meter), in what is called the "soft X-ray" region. However, such laboratory devices only work under unusual conditions and are not yet practical, as we will see in Chapter 9.

Light Is a Form of Energy

We saw earlier that light, and all other forms of electromagnetic radiation, are forms of energy. In looking at light energy, it is helpful to think of light as photons, chunks of energy. The higher the frequency or shorter the wavelength, the more energy carried by a photon.

Matter absorbs and emits photons. When it absorbs a photon, it gains energy; when it emits or radiates a photon, it releases energy. The amount of energy can be increased by increasing the photon energy, increasing the number of photons, or both. That is, you can get 10 electronvolts of energy from 10 one-electronvolt photons, one 10-electronvolt photon, 5 two-electronvolt photons, two 5-electronvolt photons, or any other combination adding to 10 eV.

As we will see later in this chapter, the real world is not quite that simple. Materials emit and absorb light most efficiently at specific wavelengths or wavelength ranges. This means that in practice you can't count on things absorbing or emitting energy at particular wavelengths—you have to take what you can get. A material may emit light energy in 10-eV chunks, but not in 1-eV chunks.

Properties of Light

The way materials absorb and emit light energy shows the particle side of light's dual personality. Certain materials emit electrons when light strikes them in a vacuum. If you change the wavelength of light, you find that the materials emit no electrons until the wavelength is shorter than a characteristic value. At shorter wavelengths, the number of electrons emitted depends on the light intensity. However, if the wavelength is longer than that threshold level, increasing the intensity won't cause the material to emit electrons. This is called the photoelectric effect, and to understand it you need to realize that light energy comes packaged in photons. As we saw before, the photon energy increases with frequency (or decreases with wavelength). It takes a certain amount of energy to free the electron from the atom, and the photon can't do the job unless it has at least that much energy. Albert Einstein first explained the photoelectric effect in 1905, laying the groundwork for quantum mechanics. (Interestingly, Einstein received the Nobel Prize in Physics for this work, not for his better-known theory of relativity.)

Another optical phenomenon reveals the wave nature of light. The basic idea is shown in *Figure 2-2*, where a bright light illuminates two parallel slits. Light that passes through the slits forms a pattern of bright and dark bands. The explanation for this seemingly mystifying effect—called "interference"—lies in the wave nature of light. When two waves come together, they add in amplitude, as shown in *Figure 2-3*. If both have their peaks and valleys at the same point (*Figure 2-3A*), the sum of the two light waves is a wave with larger amplitude, producing a bright spot by what is called "constructive interference." However, if the peak of one light wave comes at the same point as the trough of the other, the negative amplitude of one adds to the positive amplitude of the other. If the two waves have equal amplitudes, the result is a null point, with zero amplitude as shown in *Figure 2-3B*. This "destructive interference" gives the dark stripe in the interference pattern. (The energy does not vanish; it just appears somewhere else.)

How does the two-slit pattern produce the interference stripes shown in *Figure 2-2*? Light from one source illuminates the two slits, so identical light waves emerge from each of them. The distance from the slit to the screen where the stripes show is fixed, but the distance light must travel to any point on that screen depends on the angle from the slit to that point. Suppose that the central point of the screen is n light waves from each slit, and the peaks of the two waves line up with each other, to give constructive interference and a bright stripe. If you move a little to one side, it will be $n + \frac{1}{4}$ wave from one slit and $n - \frac{1}{4}$ wave from the other. That means that the two waves will come together a half wave out of phase, and the peaks of one will line up with the valleys of the other, giving a dark stripe.

Interference is a very important property of light and other

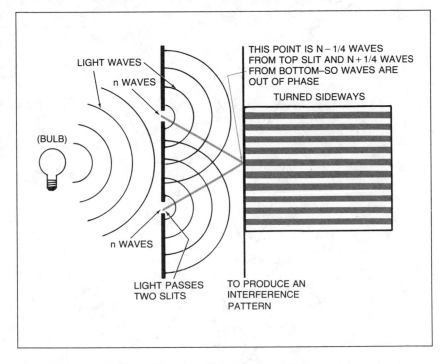

Figure 2-2. Bright light illuminating two slits causes interference.

electromagnetic waves. It occurs in many places besides the two-slit experiment. We will come upon it again and again as we learn more about lasers and optics.

QUANTUM AND CLASSICAL PHYSICS

Einstein's use of photons to explain the photoelectric effect was a key element in the development of quantum mechanics, and quantum mechanics is implicit in our understanding of lasers. Quantum mechanics differs in important ways from the "classical" physics described by Isaac Newton's laws of motion. Classical physics assumes that energy can vary smoothly, so atoms or molecules could have any amount of energy. Quantum mechanics recognizes that energy comes in discrete chunks or quanta, and that atoms, molecules, and everything else in the universe can only have discrete

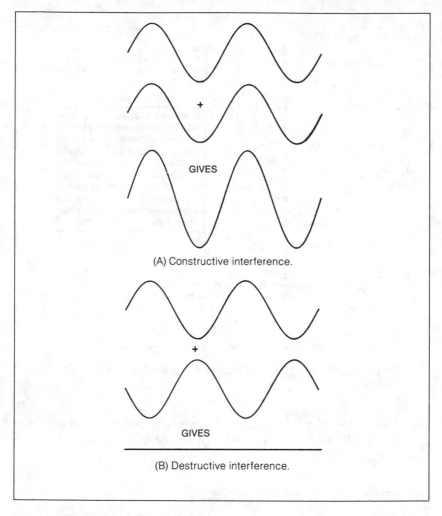

(A) Constructive interference.

(B) Destructive interference.

Figure 2-3. Addition and subtraction of light-wave amplitudes causes interference.

amounts of energy. In classical physics, energy levels can change continuously; in quantum mechanics, they change in steps. The wave picture of light is classical; the photon is quantum mechanical.

The laser is inherently a quantum mechanical device, because its operation depends on the existence of light quanta, as we will see in the next chapter. To understand the workings of the laser, we have to look at a photon-by-photon, atom-by-atom level. In particular, we must look at the quantization of energy levels within atoms. That quantization is a cornerstone both of laser physics and quantum mechanics.

ENERGY LEVELS

In our macroscopic world, energy might seem to vary continuously, like the level of sand in a pail. If we look closely at the sand, we can see that it is made up of many separate grains, and that we can add or subtract only one grain at a time. Likewise, atoms and molecules can only have certain amounts of energy. We call these energy states or levels.

For starters, let's look at the simplest atom (hydrogen) in which a single electron circles a nucleus that contains a single proton. At first glance, the hydrogen atom looks like a very simple solar system, with a single planet (the electron) orbiting a star (the proton). The force that holds the atom together is the attraction between the positive charge of the proton and the negative charge of the electron. In a real planetary system, the planet could orbit at any distance from the star. That is not true for the hydrogen atom. The electron can occupy only certain orbits, as shown in *Figure 2-4*. We show the orbits as circles for simplicity, but we can't really measure exactly what the orbits look like. Their nominal sizes actually depend on a "wavelength" assigned to the electron because matter, too, sometimes can act like a wave. The innermost orbit has a circumference of one wavelength, the next orbit a circumference of two wavelengths, and so on.

If we added energy to our simple planetary system, the planet would move farther from the star. The same happens in the hydrogen atom. As you add more energy to the atom, the electron moves to more and more distant orbits. However, there is a crucial difference between the behavior of a hydrogen atom and our imaginary planetary system. The planet can be at any distance from the sun, or could even fall into it. However, the electron can only occupy certain orbits or energy levels. These energy levels are plotted at the side of *Figure 2-4*, with labels indicating the corresponding orbits. The atom is in its lowest possible energy level—the ground state—when the electron is in the innnermost orbit, closest to the proton. (Note that the electron does not fall onto the proton.)

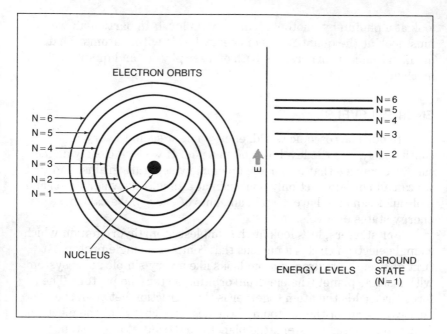

Figure 2-4. The hydrogen atom and the corresponding energy levels.

The energy levels in the hydrogen atom get closer together as they get higher above the ground state. Eventually the differences become vanishingly small. If the electron gets too much energy, it escapes from the atom altogether, a process called ionization. If we define the energy of the ionized hydrogen atom to be zero, we can write the energy of the atom E as a negative number using the simple formula:

$$E = -R / n^2$$

where,

R is a constant (2.179×10^{-18} joule),

n is the quantum number of the orbit (counting outwards, with one the innermost level).

The hydrogen atom is the standard model used to explain quantum mechanics and energy levels because its simplicity clearly shows the workings of electronic energy levels. The picture is more complicated in atoms with more electrons.

The energy level occupied by an electron can be specified as a set of quantum numbers. Physicists interpret these as identifying shell, subshell, position in a shell, spin, and other quantities. Each electron in an atom must have a unique set of quantum numbers—i.e., occupy a unique quantum state—according to the Pauli exclusion principle. These shells and subshells usually are identified by a number and letter, such as *1s* or *2p*. The number is the primary quantum number. The letter identifies subshells formed under complex quantum-mechanical rules. Each shell can contain two or more electrons, but each electron must have a unique set of several quantum numbers.

If atoms are in their ground state, electrons fill in energy levels from the bottom up—that is, from the innermost shell—until the number of electrons equals the number of protons in the nucleus. The primary quantum number of the last ground-state electron corresponds to the row the element occupies on the periodic table. *Table 2-3* shows how electronic energy levels fill up for elements with increasing numbers of electrons, for elements up through strontium (atomic number 38).

These energy levels are the same ones that play a crucial role in chemistry, determining the chemical behavior of the elements and the structure of the periodic table. The elements react with others so they can have their outermost electron shell filled. Atoms with filled outer shells, such as the rare gases neon, helium, krypton, and xenon, tend not to react with other atoms. On the other hand, atoms such as sodium, with a single outer electron, or chlorine, with an outer shell missing just one electron, are highly reactive because of their tendency to lose or gain electrons so they can have a full outer shell.

The energy levels that electrons occupy in atoms are the simplest types to describe, but there also are many more types of energy levels. Molecules, like atoms, have electronic energy levels, which increase in complexity with the number of electrons and atoms in the molecule. Atomic nuclei themselves have energy levels, although the energies involved are well above those of electrons. Molecules have energy levels that correspond to vibrations of atoms within them and rotation of the entire molecule. All these sorts of energy levels are quantized, so atoms or molecules can have only certain levels of energy.

Table 2-3. Electronic Energy Levels or Orbitals, by Increasing Energy

1s								
1s	2s							
1s	2s	2p						
1s	2s	2p	3s					
1s	2s	2p	3s	3p				
1s	2s	2p	3s	3p	4s			
1s	2s	2p	3s	3p	4s	3d		
1s	2s	2p	3s	3p	4s	3d	4p	
1s	2s	2p	3s	3p	4s	3d	4p	5s

Numbers of Electrons in Shell

2	2	6	2	6	2	10	6	2

Transitions and Spectral Lines

Nothing would happen in the universe—and we wouldn't be around to wonder why—if atoms and molecules couldn't shift from one energy level to another. The process of making changes or "transitions" between energy levels is tremendously important to laser physics. To look at the process, let's start again with the hydrogen atom.

The electron needs to gain energy to move from the ground state to a higher energy level. Conversely, it must release energy when it drops from a higher level to a lower one. One of the most convenient ways (although by no means the only way) for the electron to absorb or release energy is as a photon, a quantum of electromagnetic energy. The photon energy (the amount of energy absorbed or released) equals the difference in energy between the two energy levels.

Suppose we start with the electron in the ground state and want to raise it up one step to the first excited level of the energy-level ladder. To do so, we must give the electron exactly as much extra energy as the difference in energy between the ground level and the first excited state. Conversely, for the electron to drop from the first excited level to the ground state, it must release a photon with as much energy as the difference in energy between the two levels. In short, the photon energy equals the transition energy. The same is true for any transition—the energy of the absorbed or emitted photon equals the difference in energy between the initial and final states. Because the energy associated with each state is constant (except in certain circumstances which don't matter here), there are only a limited number of possible transition energies. Thus an electron in a hydrogen atom can emit or absorb light at only certain wavelengths.

Figure 2-5 shows the wavelengths corresponding to transitions from the ground state of the hydrogen atom to higher energy levels. The longest wavelength, 121.6 nanometers, corresponds to the 10.15-electronvolt transition energy between the ground state and first excited level. The next-longest wavelength, 102.6 nm, is the transition between ground state and second excited state. The wavelengths of transitions to higher energy levels are even shorter, but they eventually reach a limit at 91.2 nm. Light of that wavelength has enough energy to ionize the hydrogen atom, removing the electron from it completely.

The series of wavelengths in *Figure 2-5* are called the Lyman lines, after American physicist Theodore Lyman who discovered them

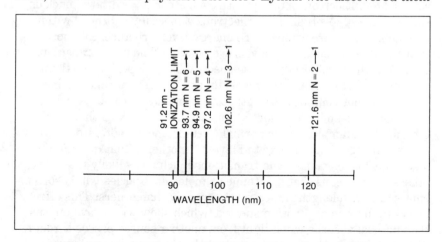

Figure 2-5. Lyman lines of hydrogen, showing energy levels involved.

in the laboratory. There are similar series of transitions at longer wavelengths involving the first, second, and subsequent excited levels. The wavelengths (λ) of all these transitions are given by the formula:

$$1 \: / \: \text{Wavelength} = \text{Constant} \: [(1 \: / \: m^2) - (1 \: / \: n^2)]$$

or

$$1 \: / \: \lambda = R \: [(1 \: / \: m^2) - (1 \: / \: n^2)]$$

where,
  m is the number of the lower energy level,
  n is the number of the higher level (the numbers increase with energy level, starting with 1 for the ground state),
  R is the Rydberg constant, 109,678 cm^{-1}.

(The strange units of "per centimeter" or inverse centimeters give the right dimensions to the answer, which comes out in centimeters. To convert it to meters, divide by 100.)

This neat ordering of energy levels is evident only in the hydrogen atom, where there is just a single electron. Add more electrons, and the energy-level picture quickly becomes more complicated. Electrons interact with each other, and with the nucleus, shifting energy levels slightly. Electrons can occupy subshells within each shell. The more complex the energy level structure, the more transitions between energy levels are possible. The more transitions, the larger the number of possible spectral lines. Superimpose them all on a single spectrum and they look almost like a set of random lines.

A further complication is that all transitions are not equally likely. One reason is that more atoms are in some states than in others. For example, under normal circumstances, more atoms are in the ground state than in excited levels. Another is that quantum-mechanical rules make some transitions much more likely than others. That means that an atomic or molecular species will absorb or emit some wavelengths much more strongly than others. This effect shows up both in absorption spectra (which show what wavelengths the material absorbs when light from another source passes through it) and emission spectra (the wavelengths the material emits when it is itself excited).

Types of Transitions

So far we have concentrated on electronic transitions, partly because we picked the hydrogen atom as our introductory example. Electronic transitions can cover a wide range of wavelengths as shown in *Table 2-4*. Those that are most important in laser physics are in the middle of the range of possibilities, involving electrons in the outer or "valence" shells of atoms or molecules. These transitions occur at ultraviolet, visible, or infrared wavelengths from about 100 nm in the ultraviolet through to near-infrared wavelengths.

Table 2-4. Representative Types of Transitions and Their Wavelengths

Transition	Wavelengths	Spectral Range
Nuclear transition	0.0005–0.1 nm	Gamma ray
Inner-shell electronic in heavy element	0.01–10 nm	X-ray
Electronic (Lyman-alpha in H)	121.6 nm	Ultraviolet
Electronic (argon-ion laser)	488 nm	Visible, green
Electronic (H, level 2–3)	656 nm	Visible, red
Electronic (neodymium laser)	1064 nm	Near infrared
Vibrational (HF laser)	2700 nm	Infrared
Vibrational (CO_2 laser)	10,600 nm	Infrared
Electronic Rydberg (H, level 18–19)	0.288 mm	Far infrared
Rotational transitions	0.1–10 mm	Far infrared to microwave
Electronic Rydberg (H, level 109–110)	6 cm	Microwave
Hyperfine transitions (Interstellar H gas)	21 cm	Microwave

Some electronic transitions can occur at longer and shorter wavelengths, but physicists usually classify those ends of the range of electronic transitions in other categories. The shortest wavelengths come from inner-shell electronic transitions in heavy elements, which involve much more energy than outer-shell transitions. These short wavelengths are considered to be X rays. On the other hand, transitions between high-lying electronic energy levels (say, levels 18 and 19 of hydrogen) involve very little energy, putting them deep in the infrared, microwave, or even radio-frequency range. Because these are qualitatively different than higher-energy transitions of outer electrons, they are put into a special class called Rydberg transitions.

Neither Rydberg transitions nor X-ray emission are likely events under normal laser conditions. Normally very few atoms or molecules are in the high-lying states needed to undergo Rydberg transitions. (Interestingly, some Rydberg lines normally not seen on earth can show up as emission lines from interstellar hydrogen, where conditions are greatly different.) Likewise, the conditions needed to break the energetic bond between inner-shell electrons and atomic nuclei are rarely met. However, it is possible to produce extremely high energy concentrations by focusing very powerful laser pulses and, as we will see in Chapter 9, development of X-ray lasers is now on the research frontier.

Transitions between nuclear energy levels can produce even higher energy photons, called gamma rays. In practice, the wavelengths of nuclear and inner-shell-electron transitions can overlap. This has led to some fuzziness in defining the boundary between X rays and gamma rays. A few researchers are trying to develop gamma-ray lasers, but they face formidible obstacles.

On the other end of the wavelength spectrum are transitions between vibrational and rotational energy levels of molecules. Vibrational transition energies typically correspond to wavelengths of a few to tens of micrometers; rotational transitions have less energy, typically corresponding to wavelengths of at least 100 micrometers.

Transitions in two or more types of energy levels can occur at once. For example, a molecule can undergo a vibrational and a rotational transition simultaneously, with the resulting wavelength close to that of the more energetic vibrational transition. Many infrared lasers emit families of closely spaced wavelengths on such vibrational-rotational transitions.

Remember in considering transitions that longer wavelengths correspond to lower energy, and shorter wavelengths correspond to higher energy. Thus, the energy of a visible transition is much larger than that of a rotational transition, even though the rotational wavelength is much larger. A combination of a rotational and vibrational transition thus has only slightly different energy than the original vibrational transition.

Transition energies or frequencies add together in a straightforward manner:

$$E_{1+2} = E_1 + E_2$$

where,

E_{1+2} is the combined transition energy,
E_1 and E_2 are the energies of the separate transition.

The same rule holds for frequencies, with ν substituted for the energy. However, wavelengths (λ) of combined transitions add by an inverse rule:

$$1 / \lambda_{1+2} = 1 / \lambda_1 + 1 / \lambda_2$$

or

$$\lambda_{1+2} = 1 / (1 / \lambda_1 + 1 / \lambda_2)$$

Stimulated and Spontaneous Emission

We have seen how an atom or molecule makes a transition to a higher energy level when it absorbs light and to a lower energy level when it releases or emits light. Light absorption can occur when a photon of the right energy is at the right place (the atom or molecule) at the right time (when the atom or molecule is in a lower energy level of that transition). It is essential that the atom or molecule be in the right energy state. For example, our simple hydrogen atom will absorb a photon with the transition energy from level 2 to level 3 only if it is in level 2, not if it is in level 1.

Emission also is a complex process, and a critical one for laser physics. Albert Einstein took the first step along the road to the laser when he realized that there could be two types of emission, spontaneous and stimulated, as shown in *Figure 2-6*. Spontaneous emission (*Figure 2-6A*) occurs all by itself; stimulated emission (*Figure 2-6B*) occurs when a photon of the right wavelength stimulates an excited atom to release energy at the same wavelength.

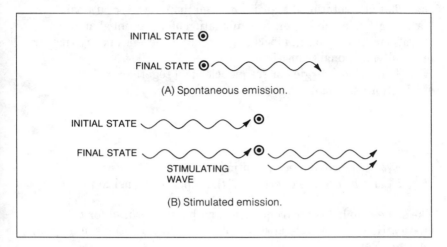

Figure 2-6. Spontaneous and stimulated emission.

Spontaneous emission occurs all by itself, and is the source of virtually all light we see in nature, from sources including the sun, stars, television sets, incandescent bulbs, and fluorescent lamps. If an atom or molecule is in an energy level above the lowest possible level (or ground state), it can drop to a lower level spontaneously, without outside intervention. It can release the excess energy as a photon of light (or in other ways, such as transferring vibrational energy to other atoms, which we won't consider here). Excited atoms or molecules have a characteristic spontaneous emission time (t_{sp}), the average time that they remain in the upper energy level before spontaneously emitting a photon and dropping to the lower level.

Einstein proposed that there could be a second type of emission, stimulated emission. Suppose you had the same excited atom, but this time illuminated it with photons having exactly the same energy as the transition the atom was going to make to the lower state. One of those photons could stimulate the excited atom to emit light on that transition. Looking at the stimulated emission as a wave, it would have precisely the same wavelength and be precisely in phase with the light wave that stimulated it.

The wave viewpoint can help show why stimulated emission has the same wavelength and phase as the light that stimulated it. The light wave stimulates the excited atom or molecule to oscillate at the light-wave frequency, which as we saw earlier corresponds the energy difference between the excited state and the lower level of the transition. That oscillation amplifies the original light wave, an effect we see as stimulated emission. The energy for the amplification comes from the atom or molecule that is stimulated, which drops to the lower level. (This viewpoint is still oversimplified from the quantum-mechanical viewpoint, but you really don't want to worry about the superposition of wavefunctions.)

We saw earlier that the word laser was coined as an acronym for *L*ight *A*mplification by the *S*timulation *E*mission of *R*adiation. We have seen that, like a laser beam, stimulated emission is all at the same wavelength and in phase—or coherent. But we haven't made a laser beam yet. It took decades for physicists to clear the crucial hurdle needed to amplify stimulated emission.

Population Inversions

The problem with stimulated emission is that it doesn't work very well under what physicists often consider "normal" conditions, thermodynamic equilibrium. At equilibrium, atoms and molecules tend to be at their lowest possible energy level. This isn't precisely the ground state, because they always have some thermal energy at temperatures above absolute zero. However, the tendency of atoms and molecules to drop to lower energy levels creates a problem in what is called population—the number of atoms or molecules at each energy level. The number decreases as the energy level increases, as shown schematically in *Figure 2-7*.

The ratio of the numbers of atoms or molecules in states 1 and 2 in thermodynamic equilibrium is given by the equation:

$$N_2 / N_1 = \exp[-(E_2 - E_1) / kT]$$

where,

N_1 and N_2 indicate the numbers of atoms in the two states,
E_1 and E_2 are the energies of the two states,
k is a constant (the Boltzmann constant),
T is the temperature in Kelvins (the absolute scale).

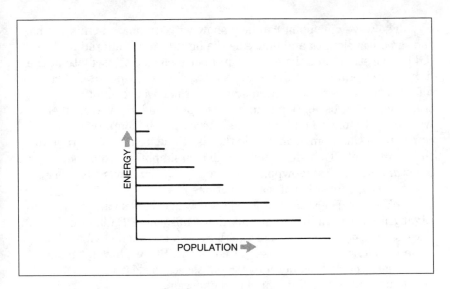

Figure 2-7. Relative population of energy levels as a function of energy above the ground state.

Under normal conditions, this ratio is quite small for transition energies corresponding to optical wavelengths. The Boltzmann constant is 1.38×10^{-23} joule/Kelvin, or 8.6×10^{-5} electronvolt/Kelvin. Plug in the numbers for room temperature (300 K) and a transition of 1 eV (slightly longer than 1000 nm) and you get a ratio of 1.5×10^{-17}. That means that in thermodynamic equilibrium, virtually all the atoms or molecules are in the ground state for a visible-wavelength transition.

Why does this make stimulated emission difficult? Let's consider a situation that at first might seem much easier: a collection of atoms with only two energy levels that are in an equilibrium population distribution when there are three times more atoms in the lower level than in the upper level. If one atom drops from the excited state and emits light spontaneously, that light could stimulate emission from another excited atom. However, if it hits an atom in the lower level first, it will be absorbed instead. The problem is that there are three atoms in the lower level for every one in the upper level. Thus, three times out of four the first atom a spontaneously emitted photon will reach will be in the lower state that can absorb it. Only a quarter of the photons encounter an excited-state atom which they can stimulate to emit light. Thus stimulated emission doesn't get anywhere, and instead we see spontaneous emission.

There is a way to make stimulated emission dominate. If more atoms are in the excited state than in the lower level, photons are more likely to stimulate emission than to be absorbed. Such a condition is called a population inversion, because it is the inverse of the normal situation where more atoms are in lower levels than in higher levels. When there is a population inversion, stimulated emission can produce a cascade of light. The first spontaneously emitted photon can stimulate the emission of more photons, and those stimulated-emission photons can likewise stimulate the emission of still more photons. As long as there are more atoms in the upper level than the lower level, stimulated emission is more likely than absorption, and the cascade of photons grows—or is amplified. Once the ground-state population becames larger, the population inversion ends, and spontaneous emission again dominates.

The problem of producing the population inversion needed for stimulated emission to dominate proved a major hang-up for physicists. Einstein's proposal, made during World War I, was considered mostly a matter of theoretical interest, because the conventional wisdom at the time said that thermal equilibrium was the normal state of matter. Anything else was unstable and wouldn't last long. The terminology that physicists used then was symbolic of that fact: they called a population inversion a "negative temperature." (The rationale came from working backwards through the Boltzmann equation, which gives a negative value for T if the higher-energy level has a larger population.)

It was not until after World War II that physicists began seriously thinking of how to produce population inversions. We'll talk more about that in the next chapter when we describe more about the workings of the laser.

INTERACTIONS OF LIGHT AND MATTER

So far, we have talked about light in fairly abstract terms. Let's take a break from abstract theory to look at another, more practical, side of the laser picture: optics and the interactions between light and matter. We can't go into extensive detail here, but it is important to have a general knowledge of these fields to understand how lasers work and how they are used.

2

We can group objects into three classes according to how they interact with light:

1. Transparent objects (e.g., glass) transmit light
2. Opaque objects (e.g., dirt or rocks) absorb light
3. Reflective objects (e.g., mirrors) reflect light

Nothing is really perfectly transparent, opaque, or reflective. Many objects transmit a little light although we think of them as opaque; hold a page of this book up to a bright light, and some light will pass through. Everything reflects some light, even objects that reflect so little they look black to us. Likewise, everything absorbs some light, even mirrors that look completely reflective. An exceptionally good solid metal mirror might reflect up to 99% of the incident light, but virtually all the rest is absorbed. Our eyes give us only a rough indication of what is highly reflective (white or shiny) and what is dark and strongly absorbing. One reason is that our perception of light or dark depends on the background. The moon looks bright against a dark sky even though it reflects only about 6% of the sunlight that strikes it.

To understand how light interacts with matter, we are going to make some simplifying assumptions. For the time being, let's ignore the fact that all materials absorb, reflect, and transmit light. Instead, we will group materials in three categories: transparent (or transmissive), absorptive (a more refined version of opaque), and reflective. We will start with the most fundamental quantity that measures the nature of transmissive materials: the refractive index.

Refractive Index

Earlier we saw that the speed of light in a vacuum, usually indicated by the letter c, is a universal constant, defined as precisely 299,792.458 kilometers per second, or roughly 300,000 kilometers or 186,000 miles per second. While the speed of light in a vacuum is a constant, the speed of light in matter—even matter as tenuous as air—is not. The ratio of the speed of light in vacuum to the speed of light in a material is the refractive index, n_{mat}:

$$n_{mat} = \frac{\text{Speed of Light in Vacuum}}{\text{Speed of Light in Material}}$$

or

$$n_{mat} = c_{vac}/c_{mat}$$

Because the speed of light is faster in a vacuum than in a material, the refractive index is greater than one. *Table 2-5* lists the refractive indexes of some common materials.

Table 2-5. Refractive Indexes of Common Materials for Wavelengths Near 500 nm

Material	Index
Air (1 atmosphere)	1.000278
Water	1.33
Magnesium fluoride	1.39
Fused Silica	1.46
Zinc crown glass	1.53
Crystal quartz	1.55
Optical glass	1.51–1.81*
Heavy flint glass	1.66
Sapphire	1.77
Diamond	2.43

* Depends on composition. Standard optical glasses have refractive indexes in this range

We learned earlier that the speed of light in a vacuum equals the wavelength times the frequency. The same holds true in other materials. Frequency remains constant, but the wavelength λ is equal to the wavelength in vacuum divided by the material's refractive index:

$$\text{Wavelength}_{mat} = \text{Wavelength}_{vacuum} / n_{mat}$$

or

$$\lambda_{mat} = \lambda_{vacuum} \ / \ n_{mat}$$

This change in wavelength affects how light travels through transparent materials.

Refraction

A ray of light bends as it passes from one transparent medium to another as shown in *Figure 2-8*. The process is called refraction and depends on the refractive indexes of the two materials. The top material has a lower refractive index, so the wavelength of light is longer in it than in the higher-index material at the bottom. As the light waves enter the higher-index material, their wavelength shrinks, but they keep in phase (each line represents a wave peak), and they oscillate at the same frequency. This bends the wavefront in the direction shown. If the light wave passes back out into the same low-index material on the other side of high-index material, the refraction simply reverses direction. You can see how if you turn the page upside down.

Note that refraction only occurs at the border between two transparent materials, not within the materials themselves. The degree of refraction is measured by measuring the angle between the direction of the light ray and what is called the "normal" angle, perpendicular to the surface of the material. If light travelling in a medium with refractive index n_1 strikes the surface of a material with index n_2 at an angle of θ_1 to the normal, the light direction in the second material is given by θ_2 in the equation:

$$n_1 \sin \theta_1 = n_2 \sin \theta_2$$

This is known as Snell's law. If the light strikes the surface at an angle θ_1, and you know the refractive indexes of the two materials, this can be rewritten to give:

$$\theta_2 = \arcsin \left(n_1 \sin \theta_1 \ / \ n_2 \right)$$

Suppose, for example, light in water ($n_1 = 1.33$) strikes the surface of crystal quartz ($n_2 = 1.55$) at 30 degrees to the normal. The formula tells us that $\theta_2 = 25.4$ degrees, meaning that it was bent closer to the normal. If the light was going in the other direction, it would be bent further from the normal. If the difference in refractive index was larger, as when going from air into quartz, the change in angle would be larger.

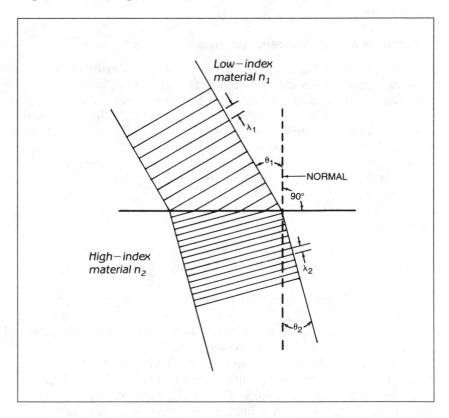

Figure 2-8. Refraction of light waves as they pass from low-index medium with refractive index n_1 to higher-index medium with index n_2.

You might wonder what would happen if the light went from a high-index material into a low-index material at a steep angle. Suppose, for example, light in a quartz crystal hit the boundary with air ($n=1$) at a 50-degree angle to the normal. Plug the numbers into the preceding equation, and you find that θ_2 equals the arcsin of 1.19.

That's impossible because the sine cannot be greater than one, and your calculator will tell you it's an error. Try an experiment, and you'll find that the light does not escape into the air; it's all reflected back into the quartz, a phemonenon called *total internal reflection* that is important in fiber optics, and some lasers and optical instruments.

Transparent and Translucent Materials

You may have been confused at some point by the distinction between transparent and translucent materials. Transparent materials are those that are both clear and transparent. Translucent materials are those that transmit light, but are cloudy, like wax paper or ground glass. The difference is that transparent materials let light pass straight through, while translucent materials scatter enough so you cannot see through it clearly. We won't deal with translucent materials in this book, and will largely avoid the question of scattering.

Reflection

Refraction is only one way to redirect light. The other is reflection. The basic law of reflection is simplicity itself:

$$\text{Angle of Incidence} = \text{Angle of Reflection}$$

This means that if light strikes a mirror surface at a 50-degree angle to the normal, it is reflected at the same angle.

There are two types of reflection, and it's important to understand the difference. Specular, or mirror-like, reflection is from a surface like a mirror or a piece of window glass that is smooth on the scale of the wavelength of light. Such a smooth surface reflects light back at the angle of incidence, just as we expect a mirror to do. Because the wavelength of visible light is very small, only 0.0004 to 0.0007 millimeter, a specular surface must be very smooth. However metal and glass surfaces can be made that smooth, and as we will see later the surfaces can be coated to enhance their reflectivity.

Diffuse reflection is from a surface that is rough on the scale of the wavelength of light. Light is reflected at the angle of incidence, but the surface is so rough that the overall effect is to scatter light in all directions. You can compare diffuse reflection to balls bouncing off a pile of rocks, and specular reflection to balls bouncing off a flat floor. The paper of this book is a good example of diffuse reflection.

All surfaces reflect some light, even those we think of as transparent because they transmit most light. You can see this at night by looking at reflections of your household lights in the glass of your windows. Special antireflection coatings can reduce these reflections.

Note that the mirrors used in high-performance optical systems are reflective on their front surfaces, unlike household mirrors, which have reflective layers behind a layer of glass. Such front-surface mirrors require more care than household mirrors, because their reflective surfaces are directly exposed to the environment. However, they are much more accurate optically, because they avoid refraction in the covering layer of glass, and the secondary reflection that can occur at the front surface of the glass.

Absorption and Opaque Materials

Perfectly opaque materials, like those that are perfectly transparent or perfectly reflective, are a convenient fiction. What actually happens is that the material absorbs a certain fraction of the light passing through a certain thickness. For example, a material might absorb 60% of the incident light each centimeter. Thus after passing through one centimeter of the material, only 40% of the original light would be left, after passing through 2 cm only 16% (0.4 × 0.4) would be left, and so on. Mathematically, we can describe this as an exponential equation:

$$I = I_0 \exp^{-ad}$$

where,
  I is the final intensity,
  I_0 is the initial intensity,
  a is the absorption coefficient,
  d is the distance travelled through the material.

This explains the well-known fact that thin slices of materials we consider opaque sometimes seem transparent. In fact, all materials have some absorption, although the value of the absorption coefficient may be very small. Likewise—at least in theory—all materials transmit some light, although when the exponential reaches values of 10^{-20} the idea of transmission becomes meaningless.

Strictly speaking, absorption is not the only effect that can block light transmission. Some light can be scattered as well, like the light scattered from dust in a sunbeam. The light is not absorbed, but instead is scattered in different directions so it is lost from the original beam. Scattering effects are described by an exponential law that looks just like the equation for absorption. In fact, the effects of scattering and absorption together are given by adding one term—s for scattering coefficient—to our equation, giving:

$$I = I_0 \exp^{-(a+s)\,d}$$

This formula is for attenuation, which includes both light absorption and scattering effects. Scattering can make important contributions to losses in laser systems, and should not be ignored.

LENSES AND SIMPLE OPTICS

The laws of reflection and refraction show how to redirect light. Refraction and reflection at flat surfaces can be useful, but they are limited. If the incoming light is made up of many parallel rays (a handy way to consider laser light), the outgoing rays remain parallel, although they may be going in a different direction. Reflection or refraction at curved surfaces can bring light rays together or "focus" them; conversely, a curved surface can also make light rays spread out. Curved reflective surfaces, like flat ones, are called mirrors. Transparent materials with smoothly polished curved surfaces are called lenses.

Lenses have one or two curved surfaces. For the moment, we will consider only the most common types of lenses, with surfaces that are flat or with a smooth surface shaped like part of a sphere. *Figure 2-9* shows the major types of lenses and how they affect a bundle of parallel light rays, such as might come from a laser. In a "positive" lens, which is thicker at the center than at the sides, refraction at front and back surfaces bends the light rays so they come together at a focal point. In a "negative" lens, which is thicker at the edges, the light rays spread out. Note that each type of lens can have different degrees of curvature on the curved surfaces.

The characteristics of lenses depend on several interrelated

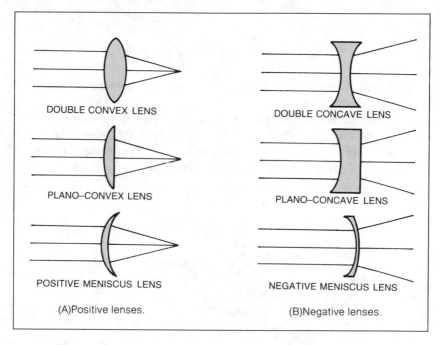

DOUBLE CONVEX LENS

PLANO–CONVEX LENS

POSITIVE MENISCUS LENS

(A)Positive lenses.

DOUBLE CONCAVE LENS

PLANO–CONCAVE LENS

NEGATIVE MENISCUS LENS

(B)Negative lenses.

Figure 2-9. The six basic types of simple lenses.

parameters, including focal length, surface curvature, size, and refractive index of the glass. Descriptions of these parameters for the different types of lenses follow.

Positive Lenses

The most important parameter for any lens is its focal length. For a positive lens, this is the distance from the lens at which it focuses parallel light rays to a point, as shown in *Figure 2-10*. The focal length depends on the refractive index of the lens and the curvature of its surfaces. Curvature here is defined as the "radius of curvature," the distance from the spherical surface of the lens to the point that is the center of the sphere of curvature. If a lens has surfaces with radii of curvature of R_1 and R_2, and is made of glass with refractive index n, its focal length f is defined by:

$$1/f = (n-1)[(1/R_1) + (1/R_2)]$$

or:

$$f = 1 / \{(n - 1)[(1 / R_1) + (1 / R_2)]\}$$

That formula is for double-convex lenses, with both surfaces curved outward.

If one surface is flat, as in a plano-convex lens, the radius of curvature is taken as infinite, so the value of that $1/R$ is zero. If a surface is concave, or curved inward, as in a positive meniscus lens, that radius of curvature is given a negative sign.

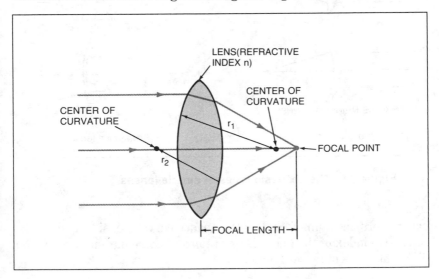

Figure 2-10. Important parameters of a positive lens.

In the real world, light rays from the same object usually are not parallel to each other. Typically they spread out, as shown in *Figure 2-11*. A positive lens also can focus these light rays, but in a different way than in parallel light rays. Such diverging rays are not all focused to the same point; instead, they form an image farther from the lens than the focal point.

The distance at which the image is formed depends on how far

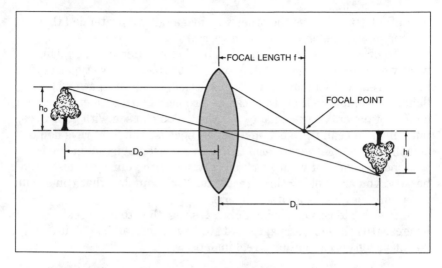

Figure 2-11. A positive lens forms a real image of an object.

the object is from the lens. If the object is a distance D_o away, the distance of the image D_i is given by:

$$1 / D_i = 1 / f - 1 / D_o$$

or

$$D_i = 1 / (1 / f - 1 / D_o)$$

The size of the image also depends on the distance. The ratio of the height of the image h_i to that of the object h_o is defined as the magnification ratio. It also depends on the distance between the lens and the object:

$$m = h_i / h_o = D_i / D_o$$

As the formula shows, the magnification is greater than one if the object is nearer to the lens than the image, meaning that the image is larger than the object. If the image is nearer to the lens than the object, the magnification is smaller than one, and the image is smaller than the object. If we look at our figure right-side up, we can see that the image is farther from the lens than the object, and thus is magnified. (It would be the other way around if we pretended that the image was the object and vice versa.)

To prevent any confusion, we should stop a minute to explain what we mean by images. Optics specialists speak of two types of images: real and virtual. Real images actually exist in space where light rays come together; they can be projected on a screen, a wall, or a piece of paper. Virtual images are those you can see with your eyes, but which you can't project on a screen because the light rays don't actually come together to form them. In a sense, a virtual image is just another view of what you normally see with your eyes, distorted because you are looking through a lens. For example, what you see in a magnifying glass is a virtual image.

Positive lenses can bring light rays together to form real images. It is those real images that we have been talking about. Negative lenses cannot form real images.

Negative Lenses

The same optical laws apply for negative lenses as for positive lenses, but they behave in somewhat different ways. The focal lengths of negative lenses are negative. You can get this result mathematically by assigning negative radii of curvature to concave lens surfaces. Its physical meaning is that they don't focus parallel light rays to a point. Pass parallel light rays through a negative lens, and they seem to spread out from a point behind the lens, as shown in *Figure 2-12*. The distance from the lens to this point is considered to be the focal length; it is given a negative value because it is on the opposite side of the lens as the focal point of a positive lens. (Optical designers have special sign conventions to keep these things straight, but we won't bother with those.) Because negative lenses don't bring light rays together, they cannot form real images, but they do form virtual images.

The equations giving magnification and image distance are not relevant for negative lenses because they apply only to real images.

Mirrors

Mirrors bend light by reflecting rather than refracting it. A flat mirror does not focus light, but it can redirect light rays. This can be useful in laser systems or other optical systems, where light must be

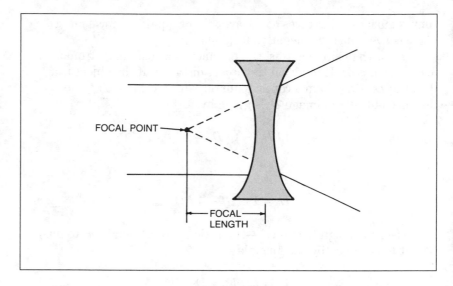

Figure 2-12. Important parameters of a negative lens.

bent around a corner, or redirected to keep it from escaping from an optical system.

Curved mirrors, like lenses, can focus parallel light rays, or make them spread apart. Mirror surfaces, like lens surfaces, can be either concave or convex. A concave mirror seems hollow, like the inside of a bowl. If the surface is spherical (as is usual for simple optics), parallel light rays striking it are focused to a point, forming a real image analogous to the one formed by a positive lens. A convex mirror is a rounded surface, like the outside of a ball. Parallel light rays striking it are spread out, as they would be by a negative lens.

Like lenses, spherical mirrors have focal lengths, which indicate the distance from the mirror at which parallel rays are focused. The focal length f is defined by the equation:

$$f = R\,/\,2$$

where,
   R is the mirror's radius of curvature.

To follow the same sign convention as for lenses, we assign a positive radius of curvature to a concave surface and a negative radius to a convex surface. This gives a concave mirror a positive focal length and a convex mirror a negative focal length.

As with positive lenses, the distance of a real image from a (concave) mirror depends on the object distance. If the object is a distance D_o away from a concave mirror with focal length f, the distance of the real image D_i is given by:

$$1 / D_i + 1 / D_o = 1 / f$$

or

$$D_i = 1 / (1/f - 1 / D_o)$$

Likewise, the magnification m—the ratio of object height h_o to image height h_i—is given by the formula

$$m = h_i / h_o = D_i / D_o$$

Look back to the formulas for a positive lens, and you will note they are the same except for a rearrangement of terms. Some optics books use sign conventions that make the terms look different (by assigning a negative sign to distances measured in one direction), but the equations are otherwise the same because the same fundamental thing is happening—light rays are being focused.

Wavelength Effects

We have simplified our discussion of optics by ignoring effects of wavelength, but we can't do that in the real world. Everything changes with wavelength. Refractive index is a function of wavelength. So are the fractions of light absorbed, transmitted, and reflected by an object. Materials that are transparent to visible light, like glass, air, and water, are opaque in parts of the infrared and ultraviolet. Conversely, some things that block visible light transmit light well at other wavelengths.

The details are far beyond the scope of this book, but wavelength effects do have major practical impacts. The variation of refractive index with wavelength means that simple lenses focus different colors of light at slightly different points, a problem called "chromatic aberration." The same variation causes a prism to bend

HOWARD W. SAMS & COMPANY

Bookmark

DEAR VALUED CUSTOMER:

Howard W. Sams & Company is dedicated to bringing you timely and authoritative books for your personal and professional library. Our goal is to provide you with excellent technical books written by the most qualified authors. You can assist us in this endeavor by checking the box next to your particular areas of interest.

We appreciate your comments and will use the information to provide you with a more comprehensive selection of titles.

Thank you,

Vice President, Book Publishing
Howard W. Sams & Company

COMPUTER TITLES:

Hardware
- ☐ Apple I40 ☐ Macintosh I01
- ☐ Commodore I10
- ☐ IBM & Compatibles I14

Business Applications
- ☐ Word Processing J01
- ☐ Data Base J04
- ☐ Spreadsheets J02

Operating Systems
- ☐ MS-DOS K05 ☐ OS/2 K10
- ☐ CP/M K01 ☐ UNIX K03

Programming Languages
- ☐ C L03 ☐ Pascal L05
- ☐ Prolog L12 ☐ Assembly L01
- ☐ BASIC L02 ☐ HyperTalk L14

Troubleshooting & Repair
- ☐ Computers S05
- ☐ Peripherals S10

Other
- ☐ Communications/Networking M03
- ☐ AI/Expert Systems T18

ELECTRONICS TITLES:
- ☐ Amateur Radio T01
- ☐ Audio T03
- ☐ Basic Electronics T20
- ☐ Basic Electricity T21
- ☐ Electronics Design T12
- ☐ Electronics Projects T04
- ☐ Satellites T09

- ☐ Instrumentation T05
- ☐ Digital Electronics T11

Troubleshooting & Repair
- ☐ Audio S11 ☐ Television S04
- ☐ VCR S01 ☐ Compact Disc S02
- ☐ Automotive S06
- ☐ Microwave Oven S03

Other interests or comments: _____

_____

_____

Name_____

Title _____

Company _____

Address _____

City _____

State/Zip _____

Daytime Telephone No. _____

A Division of Macmillan, Inc.
4300 West 62nd Street
Indianapolis, Indiana 46268

27274

Bookmark

HOWARD W. SAMS
& COMPANY

To order, return the card below, or call 1-800-428-SAMS. In Indiana call (317) 298-5699.

Please send me the books listed below.

Title	Quantity	ISBN #	Price

Subtotal _____

Standard Postage and Handling **$2.50** _____

All States Add Appropriate Sales Tax _____

TOTAL _____

Enclosed is My Check or Money Order for $ _____

Charge my Credit Card: ☐ VISA   ☐ MC   ☐ AE

Account No. _____

Expiration Date _____

☐ Please add my name to your mailing list to receive more
information on related titles.

Name (please print) _____

Company _____

City _____

State/Zip _____

Signature _____
(required for credit card purchase)

Telephone # _____

☐☐☐☐☐ ☐☐☐☐☐ ☐☐☐☐☐ ☐☐☐☐☐   ☐☐☐ ☐☐☐☐☐

27274

HOWARD W. SAMS & COMPANY

Dept. DM
4300 West 62nd Street
Indianapolis, IN 46268-2589

different wavelengths of light at different angles, so it can spread out the colors of the spectrum.

Wavelength effects also influence the selection of optical materials. Silica-based glasses and certain plastics that look transparent to the eye are fine for visible-light lenses, but you need other materials for the infrared and ultraviolet rays. In fact, you need different materials for different infrared and ultraviolet wavelengths. Some materials strongly absorb certain wavelengths but are comparatively clear at nearby wavelengths.

Further complications come from physical problems with the materials themselves. For example, salts such as sodium chloride or potassium chloride are among the most transparent materials in some parts of the infrared. However, these salts also are so soluble in water that they absorb moisture from the air. Leave them unprotected in a moist room, and eventually a very expensive lens may turn into a salty puddle! Other exotic optical materials are toxic (some contain poisonous thallium), or have other limitations.

Go far enough from the visible region, and air itself starts to become a problem. At ultraviolet wavelengths shorter than about 200 nanometers, air becomes opaque. Scientists call such wavelengths the "vacuum ultraviolet" because they have to perform experiments in a vacuum lest the air absorb the light they're using. Water vapor, carbon dioxide, and other air molecules also absorb strongly in parts of the infrared, although not as strongly as in the vacuum ultraviolet.

Optical Complexities

Optics is a complex field, which goes far beyond the realm of the simple lenses and mirrors we have described so far. If you are going to delve seriously into optics, you probably will want to get a book that deals specifically with that topic.

In this book, we treat optics as an adjunct to laser technology. That means we cover the basic elements of optics you need to know to understand lasers, but not the wealth of detail involved in optics per se. So far, we have given only a brief introduction to the field. In Chapter 5, we will talk more specifically about optics used with lasers.

2

What Have We Learned?

- Light behaves both like waves and like particles (photons).
- Electromagnetic waves can be identified by photon energy, wavelength, or frequency. Wavelength is the distance between wave peaks; frequency is the number of wave peaks passing a point per second.
- The speed of light is a universal constant, 300,000 kilometers per second. Wavelength equals the speed of light divided by frequency.
- Photon energy equals Planck's constant times frequency.
- Electromagnetic waves cover a vast range of wavelengths. Visible light has wavelengths of about 400 to 700 nm.
- Most lasers emit visible, ultraviolet, or infrared light.
- Materials emit and absorb light at characteristic wavelengths, when they make transitions between different energy levels.
- Light waves add in amplitude, causing interference effects.
- Quantum physics recognizes that energy comes in discrete chunks or quanta. In classical physics, energy levels can change continuously.
- An electron's energy level is specified by quantum numbers.
- A transition to a lower level releases energy; a transition to a higher level absorbs energy.
- Energy level structures become more complicated as more electrons are added to an atom.
- All transitions are not equally likely, because of differences in energy-level populations and transition probabilities.
- Electronic transitions can occur at microwave to X-ray wavelengths, but the most important ones are at visible, near-ultraviolet, and near-infrared wavelengths. Vibrational transitions typically occur in the mid-infrared; rotational transitions occur at far infrared or microwave wavelengths.
- Transition energies or frequencies add together simply, but an inverse addition rule is needed for wavelength.
- There are two types of emission, spontaneous and stimulated.

Most light we see in nature is spontaneous emission. Lasers produce stimulated emission.

- Stimulated emission can dominate only if more atoms occupy an upper level than a lower level, a condition called a population inversion. However, at thermodynamic equilibrium more atoms are in lower states than upper ones.

- Objects transmit, reflect, and absorb light.

- Refractive index is a fundamental quantity for all optical materials; it is the ratio of the speed of light in a vacuum to the speed of light in the material.

- Refraction bends light rays as they pass between transparent media.

- Total internal reflection occurs when light tries to leave a high-index material at a steep angle, but can't get out.

- The angle of incidence equals the angle of reflection.

- Opaque materials absorb light strongly, according to an exponential law.

- Transparent materials with smoothly polished curved surfaces are called lenses.

- The focal length of a positive lens is the distance from the lens at which it focuses parallel light rays to a point. A positive lens focuses diverging light rays from an object to form a real image. The distance and size of the image depend on the focal length of the lens and the object distance and size.

- You can see virtual images, but you can't project them on a wall.

- Negative lenses spread out parallel light rays and have a negative focal length.

- Curved mirrors can focus light by reflecting it. Concave mirrors can form real images.

- Refractive index and other material properties depend on the wavelength.

WHAT'S NEXT?

In the next chapter, we will learn about the internal workings of a generalized laser. Later on, we will learn about specific types of lasers, which can differ greatly in detail.

2

Quiz for Chapter 2

1. A carbon-dioxide laser has a nominal wavelength of 10.6 micrometers. What is its frequency?
   a. 300,000 hertz
   b. 2.8×10^{13} hertz
   c. 1.06 gigahertz
   d. 2.8×10^{10} hertz
   e. None of the above

2. What is the photon energy for an infrared wave with frequency of 10^{12} hertz?
   a. 10.6 micrometers
   b. 6.63×10^{-34} joule
   c. 6.63×10^{-22} joule
   d. 10.6×10^{22} joules
   e. About one joule

3. What is the wavelength of the infrared wave in Problem 2?:
   a. 3×10^{-4} meter
   b. 300 millimeters
   c. 300 kilometers
   d. 300 nanometers
   e. 10.6 micrometers

4. What is the metric prefix for 10^{-9}?
   a. Kilo
   b. Giga
   c. Pico
   d. Nano
   e. Femto

5. Calculate the wavelength of the transition in the hydrogen atom from the $n = 2$ energy level (the second orbit out) to the $n = 3$ level. This is the first line in the Balmer series of spectral lines
   a. 121.6 nm
   b. 91.2 nm
   c. 656 nm
   d. 632.8 nm
   e. 900 nm

6. What is the wavelength of a transition that corresponds to the combination of transitions at 500 and 700 nm?
   a. 200 nm
   b. 1200 nm
   c. 600 nm
   d. 292 nm
   e. None of the above

7. At thermodynamic equilibrium and room temperature (300 K), what is the ratio of populations at the upper and lower level of a transition with photon energy of 0.1 electronvolt? (i.e, with $E_2 - E_1 = 0.1$ eV).
   a. 0.0207
   b. -3.9
   c. 1
   d. 0.001
   e. 0.127

8. What is the ratio in problem 7 if the temperature is reduced to 100 K?
   a. 0.0207
   b. 0.27
   c. 1
   d. 0.001
   e. 0.127

9. Light in a medium with refractive index of 1.2 strikes a medium with refractive index 2.0 at an angle of 30 degrees to the normal. What is the angle of refraction (measured from the normal)?
   a. 50 degrees
   b. 60 degrees
   c. 20 degrees
   d. 17.5 degrees
   e. 15 degrees

10. A material has an absorption coefficient of 0.5 per centimeter, and its scattering coefficient is negligibly small. What fraction of incident light can pass through a 5-centimeter thickness?
    a. 0.5
    b. 0.1
    c. 0.082
    d. 0.01
    e. None, the material is opaque

11. A positive lens with a focal length of 10 centimeters forms a real image of an object 25 centimeters away from the lens. How far is the real image from the lens?
    a. 5 cm
    b. 10 cm
    c. 15 cm
    d. 20 cm
    e. 25 cm

12. What is the ratio of image size to object size for the case in Problem 11?
    a. 2
    b. 1.5
    c. 1
    d. 0.667
    e. 0.5

How Lasers Work

ABOUT THIS CHAPTER

In the last chapter, we learned the basic physics and optics needed to understand how lasers work. In this chapter, we will learn what goes on inside lasers—how they are energized and how they produce beams. Here we will learn about the general principles of laser operation; later on we will learn about laser characteristics and specific lasers.

PRODUCING POPULATION INVERSIONS

In the last chapter, we saw that you need a population inversion to make a laser. If the population distribution is at equilibrium, with more atoms or molecules in the lower level of a transition than in the upper level, absorption will soak up any stimulated emission because there are more absorbers than emitters. However, if there are more atoms or molecules in the upper level, there are more emitters than absorbers. Thus, a photon with the transition energy is likely to encounter an excited state and stimulate emission before it is absorbed.

Note that the population inversion need only be on a single transition. An atom or molecule can occupy many energy levels. Transitions can occur between any pair of levels, although in practice quantum-mechanical rules and energy considerations make some transitions much more likely than others. The minimum requirement for laser action is that at least one high level have a higher population than a lower level.

There are two basic ways to produce such a population inversion. The more obvious one is to put an excess of atoms or molecules in the higher state. However, it's also possible to depopulate the lower level, or pick a system where the lower level is unstable, so there are few or no atoms in the lower level. If the laser is to operate continuously, both population of the upper level and depopulation of the lower level are important, because accumulation of too many atoms or molecules in the lower level can end the population inversion and stop laser action.

Excitation Mechanisms

The standard way to produce a population inversion is by putting energy into the laser medium to excite atoms or molecules to high energy levels. However, it is not enough to deposit energy in a way that heats the material. Under normal conditions of "thermodynamic equilibrium," merely heating will always put more atoms into lower energy levels than in higher states. The ratio of atoms in two energy levels 1 and 2 at a given temperature T is given by:

$$N_2 / N_1 = \exp[-(E_2-E_1) / kT]$$

where,
N_1 and N_2 are the number of atoms in each level,
E_1 and E_2 are the energies of the two levels,
k is the Boltzmann constant (1.38054×10^{-23} joule per degree Kelvin).

As long as E_2 is greater than E_1, the exponential is negative, and the ratio is below one. Heating the material increases the average energy, but does not make N_2 greater than N_1.

To produce a population inversion, you must selectively excite the atoms or molecules to particular energy levels. This requires considerable ingenuity, although as we will see later in this chapter, it can occur in nature. (However, if you think only in terms of matter at thermodynamic equilibrium, the only way to make N_2 greater than N_1 in the equation is to make temperature a negative number. This is why before the advent of the maser and laser some physicists spoke of a population inversion as a "negative temperature.")

The two most common laser excitation techniques are light and electricity. Both light and electrons can transfer energy to atoms or molecules, selectively exciting them to certain higher energy levels, as described in more detail for individual lasers in Chapters 6, 7, and

8. This excitation does not have to put the atom or molecule directly in the upper level of the laser transition. It may excite the light-emitting species to a higher level, from which it can drop to the upper laser level. Or it may excite other atoms in a mixture of gases, and those other atoms can transfer the energy to the light-emitting atom or molecule. As we will see later on, the more complex-sounding approaches often make better lasers.

Metastable States and Lifetimes

These excitation techniques won't work unless atoms or molecules have the right type of energy-level structures. Normally excited states have short lifetimes and will release their excess energy by spontaneous emission very rapidly, in a matter of nanoseconds (billionths of a second). Excited states with such short lifetimes don't last long enough to be stimulated to emit their energy; most of it emerges as spontaneous emission. What is needed is a longer-lived excited state.

Such states do exist. They are called "metastable" because they are unusually stable on an atomic time scale, although they may only last for a millisecond (a thousandth of a second) or a microsecond (a millionth of a second). They are very important in laser physics because they make the best kind of upper laser level. Excited atoms and molecules can stay in an metastable excited state for long enough to produce significant amounts of stimulated emission.

On a more quantitative basis, laser action requires a build-up of population in the upper laser level. This is possible only if the upper laser is populated faster than it decays, so the population of the upper level exceeds that of the lower level. The longer the spontaneous-emission lifetime, the slower the decay rate—and the more practical it becomes to put enough atoms or molecules in the upper laser level to get laser output.

Masers and Two-Level Systems

Laser operation will make more sense if we look at some specific examples. We mentioned earlier that Charles Townes made the first maser at Columbia University several years before the first laser. His maser, like the lasers that followed, required a population inversion. However, Townes relied on an unusual trick to produce the population inversion. He used a molecular-beam technique to separate excited ammonia molecules from ground-state molecules. He threw away the ground-state molecules, but kept the excited molecules separate. Because his sample held only excited molecules, he had a population inversion and could demonstrate maser action.

That trick does not work for lasers, and it was not long before scientists found other ways to make masers. However, it played an important role in maser and laser development by proving that population inversions could be produced—something some physicists had doubted.

Three- and Four-Level Lasers

Townes's first ammonia maser involved only two energy levels: the upper and lower laser levels. Practical laser systems involve three, four, or more energy levels, depending on how the energy is transferred.

The simplest type of energy-level structure is the three-level laser, shown in *Figure 3-1*. For simplicity, we assume that all the atoms start out at the ground state, which also is the lower laser level. Then most of them are excited to a higher energy level. They quickly fall from that short-lived level to the metastable upper laser level, which has a much longer lifetime (typically a thousand times or more longer). The result is a population inversion between the metastable state and the depopulated ground state, which are the upper and lower laser levels. Because this population inversion means that more atoms are available in the upper state for stimulated emission than in the lower level for absorption, stimulated emission can dominate on the laser transition.

Maiman's ruby laser is one of the most important examples of this three-level scheme. Although the system works, it is not ideal. One problem is that the ground state is also the lower laser level. To produce a population inversion, you must excite a majority of the atoms to the upper laser level. This requires an intense burst of energy—for which Maiman used a bright flashlamp. The population inversion also is very difficult to sustain, so three-level lasers operate in pulsed mode.

Most practical lasers involve at least four energy levels, as shown in *Figure 3-2*. As in the three-level laser, the excitation energy raises the atom (or molecule) from the ground state to a short-lived highly excited level. The atom or molecule then drops quickly to a

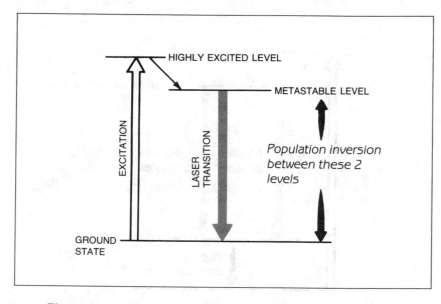

Figure 3-1. Energy levels in a three-level laser.

metastable upper laser level. The laser transition then takes the atoms or molecules to a lower state, but not all the way to the ground state in a single step. After they drop to the lower level, the atoms or molecules eventually lose the rest of their excess energy by spontaneous emission or other processes and drop to the ground state.

The key difference in three and four-level lasers is that the lower level is not the ground state in a four-level laser. Why is this so important? Because normally most atoms or molecules are in the ground state. To produce a population inversion on a laser transition in which the ground state is the lower laser level, you have to raise most of the atoms or molecules from the ground state to the upper laser level. That requires an intense burst of energy, such as the flashlamp pulses Maiman used to excite his ruby laser. Suppose, however, that normally only a few atoms or molecules were in the lower laser level—say 1%. Then you would have a population inversion if you could excite only 2% of the atoms to the upper laser level. That is much easier than raising over half the atoms to the upper laser level to produce a population inversion over the ground state.

Having the lower laser level separate from the ground state also

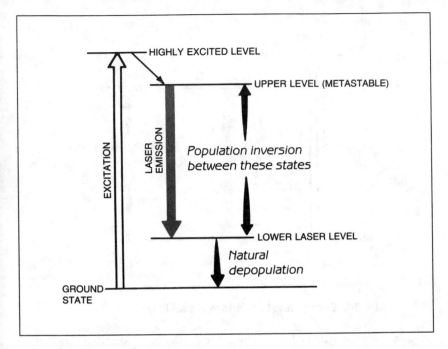

Figure 3-2. A four-level laser scheme.

brings a more subtle advantage. Atoms or molecules spontaneously drop from the lower laser level to lower levels or the ground state. If the lifetime of the lower laser level is much shorter than that of the upper level, this steady depletion of the lower level helps sustain the population inversion, by avoiding an accumulation of atoms or molecules in the ground state. This lets some lasers produce a continuous beam (often called "continuous-wave" emission). If the lower level has too long a lifetime, the population of that level can build up until it is larger than that of the upper laser level, stopping laser emission. There are some such lasers.

If you look carefully at the operation of real lasers described in later chapters, you will see that the actual energy-level structures are more complex. Excitation is not always to a single high level; it may be to a group of levels, all of which decay to the same upper laser level. That actually is good news, because it means that excitation can be over a wider range of energies than if it had to raise the atom to a single specific state. The picture at the lower end of the energy scale also can be more complex, especially where depopulation of the lower laser level is important. The upper and lower laser levels in some systems are much farther above the ground state than indicated in *Figure 3-1* or *Figure 3-2*. This makes it easier to control their populations, but it tends to limit overall operating efficiency.

We also have considered only two levels, a single upper level and a single lower level. In most lasers there are more levels. For example, there may be two closely spaced upper levels, each leading to a separate lower level. This can produce multiple transitions, which are possible in many lasers. For example, we often think of the helium-neon laser as emitting only on a single red transition. However, it also can operate on infrared transitions and on green, yellow, and orange wavelengths, if the optics are suitably designed. Other lasers, such as carbon dioxide, can simultaneously operate on many closely spaced wavelengths because the upper and lower levels are split up into many sublevels.

Another important fact is that the active media in many practical lasers contain more than one species of atoms or molecules. One species may capture the excitation energy efficiently, then transfer the energy to another species to produce a population inversion and laser action. One example is the helium-neon laser, in which helium atoms capture the excitation energy from electrons passing through the gas, and transfer that energy to neon atoms, generating a population inversion in the neon. In other lasers, another gas may be added to depopulate the lower laser level. In solid-state lasers, the light-emitting atom normally is embedded in a host crystal which doesn't generate light.

Natural Masers & Lasers

One major reason it took over four decades from Einstein's prediction of stimulated emission to the demonstration of the first laser was that physicists thought it would be very hard to produce a population inversion. They apparently didn't realize that matter could naturally be in a state other than thermal equilibrium. Ironically, it was only after people began building masers and lasers that we discovered natural masers—in outer space. Charles Townes, now at the University of California at Berkeley, has been a leader in studying such "cosmic masers."

Natural masers are gas clouds near hot stars. Starlight excites molecules in the gas to high energy levels, and the molecules drop down the energy-level ladder to a metastable state. If there is a suitable lower laser level, this can produce a population inversion, and laser action. Some scientists think similar processes might occur in the atmospheres of some planets, and that they could generate infrared stimulated emission at the same wavelengths as man-made carbon-dioxide gas lasers.

Although they produce stimulated emission in the same way as man-made masers and lasers, cosmic masers differ from them in important ways. Although they may radiate tremendous amounts of energy, cosmic masers do not produce beams; they emit in all directions, like any cloud of hot interstellar gas. Without special instruments, in fact, they look just like other gas clouds—which is what astronomers first thought they were. It was only when astronomers studied the wavelengths of light that they emit that they discovered their special nature. Cosmic masers emit strongly at transition wavelengths of certain molecules, such as carbon monoxide. Hot gas emits a broad, continuous spectrum, like the white light from a light bulb.

RESONANT CAVITIES

A population inversion is not all it takes to make a laser. A hot blob of gas with an inverted population, like a cosmic maser, emits light in every direction. The light may be stimulated emission, and it may be at a single wavelength, but it isn't concentrated in a laser beam. If our eyes were sensitive to the right wavelength, that kind of stimulated emission would look like ordinary colored light. To extract energy efficiently from a medium with a population inversion and make a laser beam, you need a resonant cavity that helps build up (or amplify) stimulated emission by feedback—reflecting some of it back into the laser medium.

To see what happens, let's start by looking at the process of amplification.

Amplification and Gain

Stimulated emission can amplify light. One photon with energy corresponding to a laser transition can stimulate the emission of a cascade of other photons at the same wavelength. If you think of that initial photon as a signal, the stimulated emission amplifies it. One of the first uses of masers was to amplify weak cosmic microwave signals, a job at which they excel because they amplify only a narrow range of frequencies and ignore the background noise. Light amplification by the stimulated emission of radiation (remember the origin of the word laser?) can produce a powerful beam of light.

The workings of laser amplification are shown in *Figure 3-3*. The first photon comes from spontaneous emission on the laser transition. When this light wave encounters an atom in the upper laser level, it stimulates it to emit light and drop to the lower laser level. If it encounters an atom in the lower laser level, it could be absorbed. (Stimulated emission and absorption are not automatic—but the probabilities are equal if the photon encounters an atom in the right states.) Remember, though, that we have a population inversion, so more atoms are in the upper laser level than in the lower level, and the light is more likely to stimulate emission than to be absorbed.

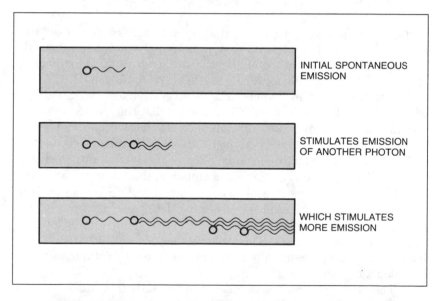

INITIAL SPONTANEOUS EMISSION

STIMULATES EMISSION OF ANOTHER PHOTON

WHICH STIMULATES MORE EMISSION

Figure 3-3. Initial spontaneously emitted photon stimulates emission of other photons, which stimulate more emission, and so on.

After the first stimulated emission, we have two photons of the same energy precisely in phase with each other. Like the original spontaneous emission photon, each of these is more likely to encounter an atom in the upper laser level than in the lower level. Thus, they, too, are likely to produce stimulated emission. So are the stimulated-emission photons they produce. You can see the trend— the number of photons produced by stimulated emission grows very quickly.

Laser physicists measure the degree of amplification as "gain," the amount of stimulated emission a photon can generate as it travels a given distance. For example, a gain of 2 per centimeter means that one photon generates two more photons each centimeter it travels. A gain of 0.05/cm means that a photon generates an average of 0.05 stimulated emission photon each centimeter it travels. The result is an amplification factor that to a crude approximation increases like compound interest:

$$\text{Output/Input} = \text{Amplification} = (1 + \text{gain})^{\text{Length}}$$

Thus for a gain of 0.05/cm, the amplification factor is (1.05) raised to the power of length (measured in centimeters). Even though this gain may sound small, the total amplification can grow rapidly; for 10 cm the amplification is 1.63, for 20 cm it's 2.65, and for 50 cm it's 11.5.

The expression for gain actually is a very crude approximation of the degree of amplification. The reason is that the gain term itself depends on how many atoms or molecules are in the upper and lower laser levels. (It also depends on the probability of stimulated emission, and other factors including density and temperature of the laser medium.) The more stimulated emission, the fewer atoms remain in the upper laser level to give more stimulated emission. The number of atoms or molecules in the upper laser level that can be stimulated to emit light drops, while the number in the lower level that can absorb light rises. This reduces the gain coefficient, and in many pulsed lasers eventually cancels it out and stops stimulated emission altogether.

The gain term can be useful in studying lasers, but it is important to understand what it means. The values given for laser "gain" generally are what is called "small-signal" gain—the gain when the signal being amplified is still weak, and there are plenty of atoms remaining in the upper laser level.

The precise increase in stimulated emission with distance travelled through a laser medium is too complex to worry about here.

The important point to remember is that the amount of amplification increases sharply with the distance light travels through the laser material. Thus, light emitted along the length of a rod of laser material, such as the one shown in *Figure 3-4*, would generate more stimulated emission than light emitted perpendicular to the rod's length. Even without mirrors, this concentrates laser emission along the length of the rod, as shown in *Figure 3-4A*. The emission is concentrated in an angle θ, defined by

$$\theta = \arcsin (Dn / 2L)$$

where,
   D is the rod diameter,
   L is its length,
   n its refractive index.

Some light is lost out the sides, but it is a small amount because it can travel (and experience gain) for only a small distance before escaping from the rod.

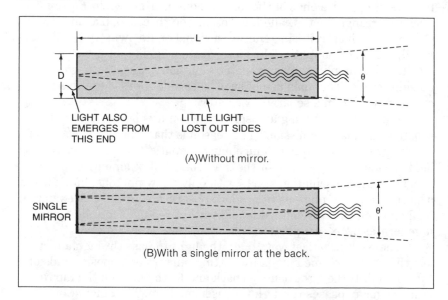

(A)Without mirror.

(B)With a single mirror at the back.

Figure 3-4. Concentration of stimulated emission in a laser rod.

A few high-gain lasers can operate in this manner, including nitrogen and excimer lasers. Typically, a mirror is placed on one end of the laser medium (*Figure 3-4B*), to reflect light back into it and stimulate more emission. However, there is no mirror on the other end of the laser medium. In such a single mirror laser, the beam-spreading angle is smaller,

$$\theta = \arcsin(Dn \, / \, 4L)$$

because of the role of the mirror. However, the beam still spreads out more rapidly than it would if the laser had mirrors on both ends. Some specialists try to avoid calling such devices lasers, but we won't worry about that fine point.

Mirrors, Laser Cavities, and Oscillation

The reason why lasers emit narrow beams *is not* because they are rod-shaped. Instead, the shape of the beam comes from the reflection of stimulated emission back and forth between a pair of mirrors at opposite ends of the laser medium, as shown in *Figure 3-5*. Such an arrangement is called an "oscillator," because, like an electronic oscillator, it can generate a signal on its own. (In the laser world, the term "amplifier" has a separate meaning—it is a laser medium that amplifies by stimulated emission a signal from some outside source. The main use of such laser amplifiers is to raise the output power from a separate oscillator.)

Why bother building an oscillator when it is not necessary to produce stimulated emission? The reason is that most laser materials have very low gain, so to produce much amplification you have to pass light through a long length of the medium. A lot of amplification is needed because oscillation starts with only a few photons, but it takes a large number of photons to produce observable power. For example, if a helium-neon laser emits one milliwatt of red light for one second, it generates 3.2×10^{15} photons.

The most practical way to get the light to pass through a long length of the laser medium is by putting mirrors on opposite ends of a tube or rod. If the light bounces back and forth between the mirrors, it makes many passes through the laser medium. The amount of stimulated emission grows on each pass through the laser medium until it reaches an equilibrium level.

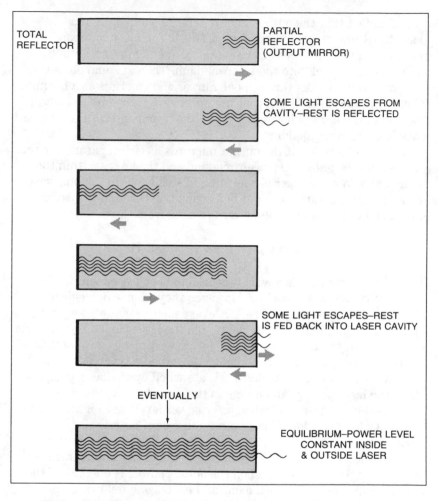

Figure 3-5. Growth of stimulated emission in a laser cavity with mirrors.

There is one important point about the laser mirrors that we haven't mentioned yet. One cavity mirror reflects essentially all of the light that reaches it, but the other reflects only some of the light back into the laser cavity. The rest emerges in the laser beam, as shown in *Figure 3-5*. If the gain is low, as in a helium-neon laser, the fraction of the light transmitted can be very low, only 1% to a few percent.

The fact that the output mirror reflects some light and transmits the rest is very important. One of the most common misunderstandings about lasers is that both end mirrors initially reflect all light back into the laser medium. Then at some point the light magically breaks through one mirror and emerges as a beam. That is not the case. The output mirror always transmits a constant fraction of the light (which emerges as the beam) and reflects the rest back into the laser medium.

The reflectivity of the output mirror is a critical parameter that depends on the gain of the laser medium and the losses within the laser cavity. When a laser is operating in steady or continuous wave mode, the amplification of light in a round trip through the cavity must balance losses and the power escaping from the cavity:

$$\text{Amplification} = \text{Loss} + \text{Output Power}$$

If the laser gain is low, it is crucial to keep losses within the laser cavity as low as possible. However, they cannot be neglected. For example, no mirror can reflect every photon that strikes it. Some light also is absorbed by material within the laser cavity. Suppose, for example, that the light is amplified by 5% as it makes a round-trip of the laser cavity, that 1% of the light is absorbed at each mirror (a total of 2%), and that an added 1% is absorbed by the laser gas. Then 2% of the laser power can emerge in the output beam.

As this example indicates, laser power levels are higher inside the cavity than outside. If the output mirror reflectivity is high, intracavity power levels can be much higher. Suppose the output mirror reflects 98% of the light back into the cavity and transmits only 2%, neglecting losses. Then the output power is only 2% of the power within the cavity. If the output beam is one milliwatt, the intracavity power is 50 mW. The difference decreases as the fraction of the light transmitted as the output beam increases, but as long as the output mirror reflects some light back into the cavity, the power inside the cavity remains higher than in the beam emerging from the laser.

Resonance

There are some subtle implications of the oscillation of light waves back and forth between mirrors in a laser cavity. One is called resonance, and depends on the wavelength of the stimulated emission and the length of the laser cavity.

To understand the nature of resonance, we need to turn back to the wave picture of light. In *Figure 3-6*, we pretend that light waves are large compared to the length of the laser cavity. (This isn't the case, of course, but you couldn't see light waves if we showed them on a realistic scale because they are so much shorter than the length of the laser cavity.) Recall, as a starting point, that stimulated emission is coherent, so all light waves are in the same phase, and that light waves add in amplitude.

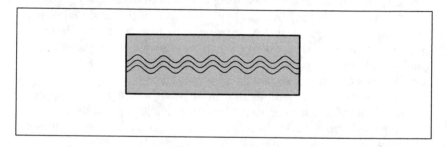

Figure 3-6. Light waves are resonant if twice the length of the laser cavity equals an integral number of wavelengths.

Figure 3-6 shows what happens when twice the cavity length is an integral multiple of the wavelength (in this case, the cavity is seven wavelengths long). Because all the light waves in the cavity are coherent, they all are in phase. Thus, each of these waves is at the same phase when it reflects from one of the cavity mirrors. For example, if light starts out at a wave peak when it is reflected from the output mirror, it will travel an integral number of wavelengths before it reaches the output mirror again, where it again will be at a peak. So will the light waves stimulated by that wave. Thus all of them will add in amplitude, by constructive interference.

Suppose, however, that twice the cavity length is not an integral multiple of the wavelength. Then each wave will reach the cavity mirror at a different phase. The waves will add in amplitude, but because they are not in phase, destructive interference will cancel them out.

The result is resonance: light waves are amplified strongly if they meet a resonance condition, and satisfy the equation:

$$N\lambda = 2 \times \text{Cavity Length}$$

where,
  N is an integer,
  λ is the wavelength.

If they do not satisfy this equation, they do not oscillate in the laser cavity.

This might seem to be a very restrictive condition which could make it hard to build a cavity that would be resonant at a particular wavelength. However, several effects combine to make laser transitions "spread out" over a range of wavelengths, and make such resonant cavities practical. Laser transitions have what is called a "gain bandwidth," a range of wavelengths over which they can exhibit gain. This turns out to be much broader than the range of wavelengths that are resonant in a laser cavity.

Furthermore, remember that light wavelengths are very much smaller than most laser cavities. For example, 30 centimeters, a typical round-trip distance in a small helium-neon laser, equals about 475,000 wavelengths of the laser's 632.8-nanometer red light. Resonance is possible not just at 475,000 waves, but also at 475,001, 475,002, etc. The wavelengths that correspond to those numbers of waves are very close, and can fit under the gain curve for the helium-neon laser, as shown in *Figure 3-7*. Each resonance value of N is said to be a "longitudinal mode" of the laser.

PRODUCTION OF LASER BEAMS

We learned that part of the light in the laser cavity emerges through the output mirror as the laser beam, but we have glossed over details of producing the beam. Beam characteristics such as size, light distribution, and the rate of spreading or "divergence" depend on design of the laser cavity and the output optics. Before we look closely at what influences the beam characteristics, we should explain a few laser concepts, including intensity distribution, oscillation modes, and beam divergence.

We may think of a laser beam simply as a collection of parallel light rays that project a bright spot on a screen. However, if we drew a line across the beam and measured the light intensity at different points, we would find that the intensity varies. Typically the beam is

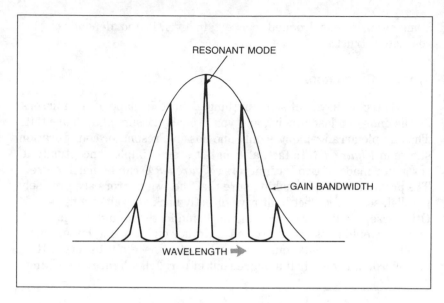

Figure 3-7. Several cavity resonances can fit within the gain bandwidth of a typical gas laser.

brightest in the middle, with intensity dropping off gradually at the sides. It can be hard to define precisely where the beam stops, so normally a cutoff is defined at a certain fraction of the central intensity (often $1/e^2$, where e is the root of natural logarithms).

We saw before that lasers oscillate in different longitudinal modes, corresponding to wavelengths which satisfy the equation:

$$N\lambda = 2 \times \text{Cavity Length}$$

where,

N is an integer,
λ is the wavelength.

Lasers also can oscillate in different transverse modes, which manifest themselves in different beam intensity patterns. Those, too, depend on resonator design, and we will look at them in more detail later.

The beam divergence also is a critical parameter. This is the angle that measures how rapidly the beam is spreading far from the laser. As we will see later, it lets you calculate size of a laser spot. Normally it, like beam diameter, is measured to points where the beam intensity has dropped to a certain level. It, too, depends on resonator structure.

Types of Resonators

So far we have not said anything about the shape of the mirrors on the ends of a laser cavity, and you probably assumed both are flat. That simple arrangement is only one possible resonator configuration shown in *Figure 3-8*. In fact, although it seems simple conceptually, a resonator made of two flat mirrors (*Figure 3-8A*) can be hard to use. The problem comes in making sure that the two mirrors are precisely parallel, so they reflect light rays directly back at each other. Otherwise, the light rays passing back and forth through the laser medium would miss the mirror on the other end, causing losses that could stop laser oscillation. The misalignment need not be large. If one mirror was only half a degree out of parallel, a light ray striking

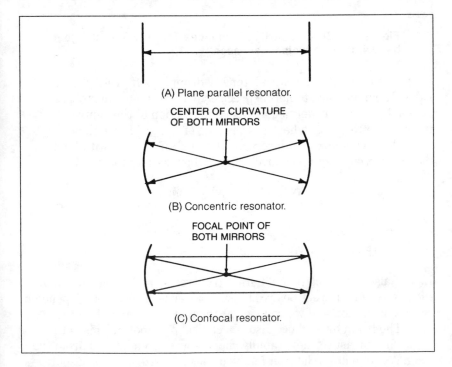

(A) Plane parallel resonator.

CENTER OF CURVATURE
OF BOTH MIRRORS

(B) Concentric resonator.

FOCAL POINT OF
BOTH MIRRORS

(C) Confocal resonator.

Figure 3-8. Major laser

it from the center of the other mirror in a 15-cm long laser would be reflected to a point 1.3 millimeters from the center of the other mirror—and would miss the edge of a cavity mirror 2 mm in diameter. Successive reflections would magnify the effects of misalignment, and increase the amount of light lost out the sides.

These light-leakage losses can be avoided if one or both of the cavity mirrors are curved, as in most of the designs in *Figure 3-8*. Follow the light rays in *Figure 3-8B, C, D, and E,* and you can see how the focusing power of the curved mirror keeps light within the laser medium. If the light is not emitted precisely along the axis of the tube, but does hit the curved mirror, the mirror often can reflect it back into the cavity. Likewise, misalignment of the mirrors doesn't matter, because their curvature automatically focuses the light back toward the other mirror. Such a configuration is called a stable resonator because light rays reflected from one mirror to the other will keep bouncing back and forth indefinitely (neglecting losses).

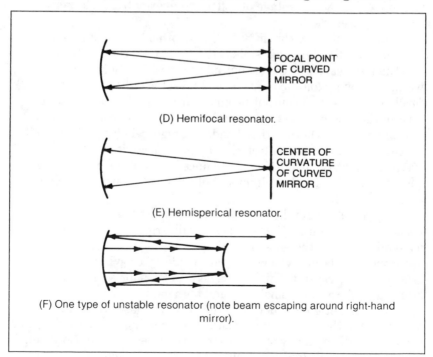

(D) Hemifocal resonator.

(E) Hemisperical resonator.

(F) One type of unstable resonator (note beam escaping around right-hand mirror).

resonator configurations.

Stable resonators are most attractive if the laser gain is low, making it vital to keep any stimulated emission from getting away.

On the other hand, if laser gain is high, it is not a problem if some light leaks out the sides of the cavity after repeated reflections. To get the maximum output power, you want to make sure you extract energy from all the excited medium between the cavity mirrors. Look carefully at *Figure 3-8*, and you can see that some stable resonators do not pass light rays through the entire volume of the excited medium. If the light rays do not go through a region, they cannot collect the energy stored there by stimulating emission.

One solution to this problem is a different type of resonator, called the "unstable resonator," which also is shown in *Figure 3-8F*. An unstable resonator gets its name because light rays successively reflected from its mirrors eventually drift out to the sides of the laser mirrors. This might make it seem undesirable, and indeed it is for low-gain lasers. However, in high-gain lasers such losses are more than offset by the advantage of collecting energy from a larger volume of laser medium.

You can also see the difference in another way. In a low-gain laser, most of the light that strikes the output mirror is reflected back into the laser cavity. If the output mirror is 95% reflective, light within the cavity (on the average) will be reflected 19 times before it finally emerges in the output beam. That makes it important to be certain that the ray can bounce back and forth between the mirrors that many times. On the other hand, in a high-gain laser, the output mirror has much lower reflectivity. If the reflectivity is 50%, on the average a light ray will be reflected once before it emerges in the output beam. This makes alignment of the cavity mirrors much less critical.

A true plane-parallel mirror cavity is neither stable nor unstable, because it neither focuses nor diverges light rays. Their main advantage is the ease of making them, especially for semiconductor lasers, where they are made by cleaving the semiconductor crystal. This causes no problems for semiconductor lasers because of their small size and high gain.

We will not go into detail on the theory of laser resonators. It bogs down in the sort of mathematical complexity you're reading this book to avoid, and many of the results are at best arcane. However, laser resonators do (quite literally) shape both the intensity distribution in the output beam and the rate at which the beam diverges.

Intensity Distribution and Transverse Modes

We earlier saw how the length of the laser cavity and the wavelength interact to establish longitudinal modes—standing waves along the length of the laser. The "transverse mode" determines the intensity distribution across the width of the laser beam. The shapes of possible transverse modes are restricted because they have to meet so-called boundary conditions. For example, the amplitude of a transverse mode must be zero at the edge of the beam. It can have one, two, or more peaks in the middle.

The simplest, or "lowest order" transverse mode is the smooth beam profile with a peak in the middle shown in *Figure 3-9*. Its shape is that of a mathematical curve called a Gaussian curve, after the famed mathematician Karl Gauss. This intensity distribution $I(r)$ as a function of distance from the center of the laser beam (r) is given by

$$I(r) = (2P \ / \ \pi d^2) * \exp(- \ 2r^2 \ / \ d^2)$$

where,

P is beam power,

d is spot size measured to the $1/e^2$ points.

This first-order mode is called the TEM_{00} mode, where the T, E, and M stand for Transverse, Electric, and Magnetic modes, respectively. The subscript numbers indicate how many zero intensity points there are between walls in the E and M directions, which are perpendicular to each other. Conventionally, the first number is the E mode, and the second the M mode.

As you might expect, there are a large family of TEM_{mn} modes, where m and n are integers. The m indicates the number of zero points or minima between the edges of the beam in one direction, and n indicates the number of minima between the edges of the beam in the perpendicular direction. Thus a TEM_{01} beam has a single minimum dividing the beam into two bright spots. A TEM_{11} beam has two perpendicular minima (one in each direction) dividing the beam into four quadrants, and so forth. The TEM_{00} mode is desirable because the spreading it experiences from diffraction approaches a theoretical minimum value. A stable resonator and a low-gain laser

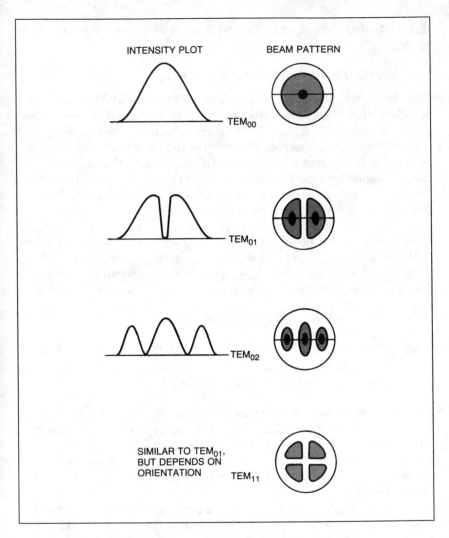

Figure 3-9. TEM$_{00}$ mode beam and a sampling of other transverse beam modes.

medium can readily produce this lowest-order mode; proper design adds losses to suppress the oscillation of higher-order modes. However, some stable-resonator lasers do operate in one or more higher-order modes, especially when they are designed to maximize output power.

Unstable resonators have fundamentally different mode structures. You can see the basic reason why in *Figure 3-10*, which examines our unstable-resonator example from *Figure 3-8F* in more detail. In this simple type of unstable resonator, the output mirror is a solid metal mirror that obstructs only part of the light emerging from the laser medium. The beam escapes around it. Thus, near the laser the beam profile has a "doughnut" cross-section—a bright ring surrounding a dark circle where the light was blocked by the mirror. If you go far enough from the laser, this intensity distribution averages out to a more uniform pattern.

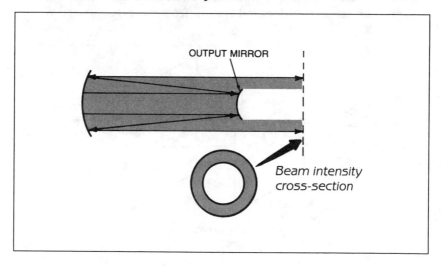

Figure 3-10. Output pattern from an unstable resonator.

The details of unstable-resonator modes are even more complex than those of stable-resonator modes, but we can safely skip their many variations. The details of mode structure generally have little practical impact. In Chapter 4 we will learn how transverse modes affect the properties of laser beams.

Beam Divergence and the Diffraction Limit

The oscillation of light back and forth in a resonant laser cavity forms a narrow beam. The divergence of that beam depends on the nature of the resonator, the size of the output aperture, and a process called "diffraction" which sets a theoretical lower limit on divergence.

Diffraction is a consequence of the wave nature of light that we haven't discussed before (although it can be considered an interference effect). It is the spreading out or scattering of light waves as they pass a sharp edge. If we were to consider light waves just as rays, diffraction would seem to bend them as they pass the edge. A better way to understand diffraction by an opening such as the one in *Figure 3-11* is to pretend that each point in the opening is emitting a new light wave, which spreads out in all directions. Interference effects concentrate the light in the forward direction, but some is scattered in a regular pattern at other angles. In *Figure 3-11*, you can see what are called Airy diffraction rings, formed when

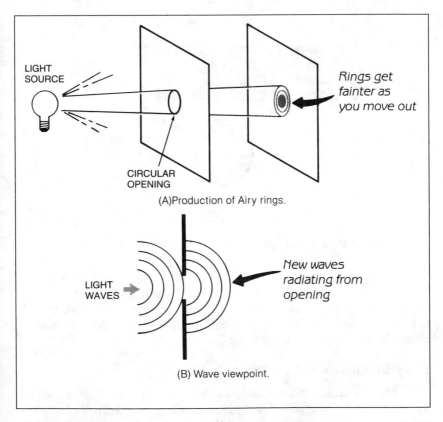

(A)Production of Airy rings.

(B) Wave viewpoint.

Figure 3-11. Diffraction of waves, and production of Airy diffraction rings.

light waves pass through a small circular hole. The central bright spot is surrounded by a series of increasingly faint rings.

Our example of light passing through a hole is similar to the way a laser emits a beam of light, if we think of the beam emerging from the output mirror as the hole. As that implies, a laser beam is subject to the same diffraction effects that spread out ordinary light. Indeed, these diffraction effects determine the minimum divergence and spot size of a laser beam. For some lasers, the divergence is larger, but for TEM_{00} beams the divergence usually is "diffraction limited."

The nominal diffraction-limited beam divergence is given by the formula:

Divergence (in radians) = Constant × Wavelength / Diameter

or

Divergence (in degrees) = $\arcsin(K\lambda / D)$

where,
 λ is the laser wavelength,
 D is the beam diameter,
 K is a constant with a value near one that depends on the intensity distribution in the beam.

You can also use a modified version of this formula to get the radius (not the diameter) of the laser spot at a large distance L from the laser:

Beam Radius = $L \times (K\lambda / D)$

(Note that this approximation is not valid for small distances.)

These simple formulas carry a very important message about laser beam size: it increases with wavelength and decreases as beam diameter increases. In practice, it means that you can't have both a tiny beam diameter and narrow divergence. If you start with a one-millimeter diameter beam from a helium-neon laser, for example, its minimum divergence is about 0.6 milliradian (mrad). Shrink the beam diameter (at the laser output mirror) to 0.3 mm, and the beam divergence rises to about 2 mrad.

If the laser beam passes through an optical system, the value of D is the diameter of the last element through which the beam passes. Thus, if you want to focus a laser beam down to a tiny spot a long distance away, the best way to do so is by focusing it through a large aperture. For example, focusing the 1-mm laser beam in the previous example through a 1-m mirror would reduce its divergence by a factor of 1000 to 0.6 microradian. Aim that beam at a target 1000 kilometers away, and you'd get a spot just 1.2 meters across. On the other hand, the beam from the 1-mm laser aperture would be 1.2 kilometers across at the same distance.

We have avoided giving a fixed value for K because it changes with the distribution of light across the emitting area or laser beam. If the light is equally bright across the entire opening, its value is 1.22. However, the value is smaller if the intensity drops off smoothly with distance from the center of the beam, as in a Gaussian TEM_{00} beam. For a Gaussian beam, the radius of the diffraction-limited spot (to the $1/e^2$ intensity point) is given by a formula that includes the radius of the beam "waist" w_0—the narrowest point in the beam inside the laser tube:

$$\text{Spot Radius} = \lambda L / \pi w_0$$

You can see that this is comparable to our earlier formula for diffraction-limited spot size, with K replaced by a smaller number.

LASER EXCITATION TECHNIQUES

Our general description of laser physics has given you an idea of how lasers work, but we have glossed over the actual excitation of atoms or molecules to produce a population inversion. The details depend on the type of laser, and will be covered in Chapters 6 through 9. However, it will be helpful first to get a general understanding of excitation processes.

Optical Pumping

Optical pumping is the use of photons to excite atoms or molecules. It was the first approach demonstrated, and conceptually it may be the simplest. Remember the simple three- and four-level lasers we described earlier. You can pump them optically by illuminating the laser material with light at the right wavelength to raise the lasing species (atom or molecule) from the ground state to the topmost level. Then the species will drop to the metastable upper laser level, and from there emit light.

If you go back and look at *Figures 3-1 and 3-2*, they make it seem that the pump light must be of a specific wavelength, because they show narrow transitions between isolated energy levels. Fortunately, this usually is not the case. The lower and— particularly—the upper levels of the pump transition usually span a range of energies. Typically, there are multiple upper levels, which all decay to the metastable upper laser level. This means that the species can absorb light at wavelengths corresponding to any transition between the ground state and those many upper levels. Because of this, many lasers can be optically pumped with a light source emitting a broad range of wavelengths, like a flashlamp. (Note, however, that some lasers may require narrow-line pumping with another laser, which can limit their usefulness.)

One other fact you should note is that the pump photon must have higher energy—or equivalently shorter wavelength—than the emitted light. This is because the light-emitting atom or molecule must be raised above the upper laser level from a level equal to or lower than the lower laser level. As we will see later, this is one of many limits on laser efficiency.

Optical pumping can be used with any laser medium that is transparent to the pump light. In practice, it is used most for solid-state crystalline lasers and liquid tunable dye lasers. The optical pumping energy may be delivered steadily or in pulses, although many optically pumped lasers cannot produce a steady beam.

Electrical Pumping

Most of the artificial light sources in our lives get their energy from electricity, and lasers are no exception. Electricity is a convenient form in which to transmit energy and, as we shall see, it also is a useful form to use in exciting lasers. (Electricity also provides energy for the light sources used in optical pumping, but here we're talking about directly exciting the laser medium with electricity.)

A fluorescent light is a useful starting point for describing electrically excited gas lasers. An electric current flows through the gas in a fluorescent tube, transferring energy to the atoms and molecules in the gas. Those excited atoms and molecules release that excess energy as light. Ordinary mercury fluorescent tubes actually emit ultraviolet light that a phosphor coating on the tube converts to visible light.

In a typical small gas laser, an electric current flowing through the gas excites atoms and molecules, raising them to the upper level in the three- and four-level laser schemes. As in fluorescent tubes, a high voltage pulse initially ionizes the laser gas so it conducts electricity, but a much lower voltage can sustain the current and power the laser.

Some electrically driven gas lasers produce steady or "continuous wave" beams as a constant current passes through the gas. Others produce pulses of light after intense electrical pulses pass through the gas. Some high-power lasers are excited by beams of electrons fired into the gas.

Electrical pumping can only be used with laser materials that can conduct electricity without destroying laser action. In practice, this is limited to gases and semiconductors.

Semiconductor Laser Pumping

Electrical currents also power semiconductor lasers, but their operation is so different from that of other electrically pumped lasers that they deserve separate mention here as well as the more detailed description in Chapter 8. The current passing through a semiconductor laser produces a population inversion in a narrow region called the "junction," where the composition of the semiconductor changes. The voltage levels needed are very small compared to those of gas lasers.

Strictly speaking, what are excited in a semiconductor laser are current carriers, electrons and vacancies called (appropriately enough) "holes" which could accommodate electrons. The current removes electrons from the crystalline lattice to produce electron-hole pairs, which generate laser light when they "recombine," with the electron dropping back down into the hole to form a lower energy state.

Other Energy Transfer Mechanisms

There are many other variations on energy transfer. Some simply involve the capture of electrical or optical energy by one species that transfers it to another. For example, in the helium-neon laser, helium atoms capture energy from an electric discharge in the gas. However, it is not the helium that lases. The helium atoms have some energy levels close enough to those of neon that they can transfer the excitation energy to neon atoms. Then the neon atoms drop into a metastable upper laser level to produce the population inversion needed to generate light.

There also are pumping techniques quite different from optical and electrical pumping. For example, chemical reactions generate excited species in "chemical" lasers. Atoms and molecules can capture energy produced in nuclear reactions to make nuclear lasers. Free-electron lasers get their energy from a beam of electrons passing through an array of magnets. We will learn more about these unusual types of lasers in Chapter 9.

Energy Efficiency

In describing optically pumped lasers, we pointed out that the pump wavelength is always less than the laser wavelength. This is symptomatic of something true for all lasers: you get less energy out in the beam than you put into the laser. In short, lasers are inherently inefficient. You must give up some energy to convert your initial energy into the more orderly form of laser light.

The first energy limitation is the simplest to see in the diagram of energy levels in a three- or four-level laser. The pump energy raises the laser species from the bottom of the energy scale to the top. The laser transition releases only part of that energy. The rest of that energy is lost in other forms. In some cases, the inherent losses from the energy level structure are low. For example, an optically pumped laser may require pump light at 500 nanometers to produce laser output at 1000 nanometers. That means that at least half of the input energy does not emerge in the laser beam because the longer-wavelength photons have less energy. However, in other systems, the laser transition is far above the ground state, so much energy is used just to raise the laser species to a level high enough to emit laser light. As we will see in Chapter 6, this effect puts a very low ceiling on efficiency of the argon-ion laser.

Another serious limitation is the fact that excitation is never 100% efficient. For example, some light used in optical pumping is not absorbed by the laser species, and some energy in an electric current is not absorbed in a gas laser.

There also are more subtle limitations. For example, laser action requires a population inversion, so some energy is put into the laser material to generate that inversion. Yet once more atoms or molecules accumulate in the lower level, the inversion no longer exists, but some atoms or molecules remain in the higher energy level, and the energy used to put them there is lost.

These effects combine to limit the overall efficiency of lasers. Some semiconductor and carbon-dioxide lasers can convert 10% or more of the input energy into output light, but more typical lasers have overall efficiency (sometimes called "wall-plug" efficiency when measured as power in vs laser light out) of 1% or less. We will look at these limitations in more detail in Chapter 4.

LINE SELECTION AND TUNING

So far, we have said little about the factors that determine what wavelengths a laser emits. As we will see later, some lasers emit only on a single transition, while others emit on two or more. Some types of lasers, such as the helium-neon laser, can emit on any of several lines, depending upon how they are designed. Several processes are involved in selecting which transitions dominate in any given laser, and several criteria must be met before laser action can occur on any transition.

Atoms and molecules have complex energy-level structures. In general, the more electrons, the more electronic energy levels are possible. Likewise, the larger a molecule, the more vibrational and rotational energy levels are possible. It might seem that the more energy levels, the more transitions are possible. However, quantum-mechanical rules make some transitions much more likely than others.

It takes several transitions with the right set of properties to make a good laser, as shown in *Figure 3-12*. These transition characteristics are inherent in the laser species.

1. A transition with high probability of excitation is needed to raise the atom or molecule to a highly excited level

2. A high-probability transition that drops the atom or molecule into the upper laser level of the population inversion

3. A metastable state with low probability of transition to lower energy levels, so atoms and molecules will remain in this state

(as a population inversion) long enough to be stimulated to emit laser light.

4. (Ideally) A high-probability transition from the lower laser level to a lower state, depopulating the lower laser level and thus maintaining the population inversion.

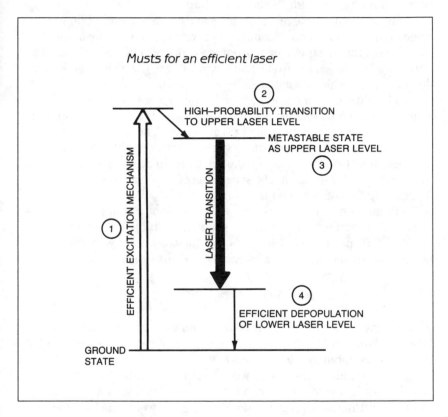

Figure 3-12. Key transition requirements for a laser.

Meeting these basic requirements is not enough to make a laser, however. The first transition in the sequence works only if there is a practical and efficient way to transfer the excitation energy and produce a population inversion. The energy can be transferred by photons, electrons, collisions with other species in a gas, or other processes—but it must be transferred efficiently enough to generate a population inversion.

The efficiency of energy transfer and the probability of transitions depend on many factors. In a gas laser, for example, these include pressure of the laser gas, concentration of other gases in the laser mixture, temperature, size of the laser tube, and density of the current flowing through the laser medium.

Because energy level structures are complex, excitation of a laser medium can produce population inversions on two or more transitions, involving one or more metastable levels. This can lead to stimulated emission at two or more wavelengths, which often is not desirable, but which usually can be controlled.

The nature of the laser amplification process provides some control over the wavelengths present, because it leads to domination of the emission by the most likely transitions. The differences in likelihood of emission need not be great. Even a slight difference in gain—such as between 0.04 and 0.05 per centimeter—is enough to lead to a dramatic difference in the strengths of two transitions originating from the same metastable level.

To understand how this works, let's look at a simple numerical example: two transitions from the same metastable level with gains 0.04 and 0.05 per centimeter. Let's assume they start with the same power, and to simplify our calculations, let's use the simple formula:

$$\text{Power} = (1 + \text{Gain})^{\text{Length}}$$

which neglects the effects of resonator mirrors, losses within the laser medium, and limits on possible gain. The relative powers on the two lines are tabulated in *Table 3-1*.

In a real laser, the power would not be amplified as much as the calculations in *Table 3-1* show, but the dependence of power on gain would result in the strong line dominating the weak one by a factor far larger than you might at first expect.

Cavity Optics

The optics of the laser cavity also can help select the wavelength at which the laser oscillates, although not in the way you might at first think. As long as the laser cavity is much longer than the wavelength, there is no problem finding a wavelength (λ) that meets

Table 3-1. Effect of Gain on Relative Power

Distance (cm)	Relative Power Weak Line	Strong Line	Ratio
0.0	1	1	1
1.0	1.04	1.05	1.01
5.0	1.22	1.28	1.05
10.	1.48	1.62	1.09
100.	50.5	131.	2.59
1000.	1.08×10^{17}	1.55×10^{21}	14351.

the criteria $N\lambda = 2 \times$ Length. Laser transitions occur over a range of wavelengths wide enough that some wavelength within that band is sure to meet the oscillation criteria.

However, laser cavity optics can suppress transitions by increasing the loss at the corresponding wavelengths. For example, suppose a laser can emit at two transitions, 400 and 600 nanometers, but you only want it to oscillate at 400 nm. You could use mirrors with different reflectivities at the two wavelengths. If the mirrors reflected light at 400 nm, but transmitted 600-nm light, there would be higher cavity losses and no feedback at the longer wavelength, so the laser would only generate the shorter wavelength—even if the laser medium had the same gain at both wavelengths.

It's also possible to tune laser emission over a range of wavelengths by putting a prism or diffraction grating into the laser cavity, as shown in *Figure 3-13*. The prism or grating spreads out the wavelengths of light at a range of angles. The cavity also contains a mirror which reflects that spread-out light. However, only one wavelength is bent at the proper angle to oscillate back and forth between the mirrors in the laser cavity. Shorter and longer wavelengths go off to the sides of the cavity and are lost. As a result, the net gain in the laser cavity is highest at the wavelength selected by the prism or grating. The output wavelength can be tuned across a

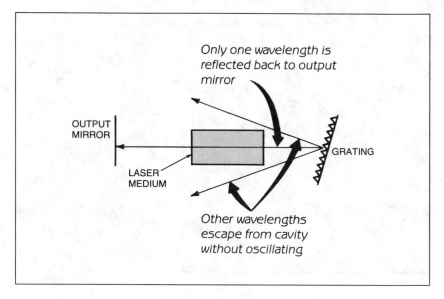

Figure 3-13. Arrangement for tuning laser wavelength continuously.

range of wavelengths (usually limited by the laser material) by turning the grating, prism or mirror. This arrangement is used in tunable dye lasers.

What Have We Learned?

- If the population is inverted, stimulated emission is more likely than absorption.

- Both populating the upper level and depopulating the lower level are important to maintain a population inversion. The upper laser level must have an unusually long lifetime.

- To produce a population inversion, you must excite specific states, not merely heat the material.

- The two most common laser excitation techniques are light and electricity.

- Practical lasers involve three or four energy levels. The three-level laser is not ideal because the ground state is the lower laser level, and thus must be depopulated to produce a population inversion. A four-level laser works better because it depopulates the lower laser level better.

- Many laser media contain more than one atomic or molecular species.

- Cosmic lasers emit light in all directions; they do not generate beams.

- A resonant cavity extracts energy efficiently from a medium with a population inversion.

- Gain measures growth of stimulated emission as light passes through the laser medium. Saturation keeps laser gain from reaching the "small-signal" value that assumes there are no limits on the growth of laser emission.

- Stimulated emission is concentrated along the length of the laser medium even without mirrors in a high-gain medium.

- Oscillation of light between mirrors at the opposite ends of the laser cavity gives the laser many of its distinctive properties.

- Oscillator cavities increase the effective length of the laser medium, producing reasonable powers from low-gain materials.

- The output mirror transmits some light to form the laser beam.

- Round-trip amplification equals loss plus output power in the laser beam.

- Laser power levels are higher inside the cavity than outside.

The smaller the output coupling, the higher the intracavity power.

- Light waves are resonant in a laser cavity if twice the cavity length is an integral number of wavelengths, and their amplitudes add by constructive interference.

- Beam characteristics depend on design of the laser cavity and output optics.

- Curved resonator mirrors help confine laser light in the cavity. They work best for low-gain lasers.

- Unstable resonators let light leak around the output mirror; they work best for high-gain lasers.

- Transverse modes are the pattern of light intensity across the laser beam. TEM_{00} is the lowest transverse mode, with intensity rising smoothly to a peak in the middle of the beam. There are a large family of TEM_{mn} modes.

- Beam divergence depends on the size of the output aperture and the laser wavelength.

- Pump light must have shorter wavelength than the laser transition.

- Semiconductor lasers are powered by currents passing through a p-n junction.

- Lasers are inherently inefficient, and suffer from several inevitable loss mechanisms. Laser excitation is never 100% efficient.

- Laser excitation can produce population inversions on multiple transitions. Laser amplification makes the strongest transitions dominant. Cavity optics can suppress oscillations by increasing losses at certain wavelengths.

- A wavelength-selective element in the cavity can tune output wavelength.

WHAT'S NEXT

Now that we have looked at the basic mechanisms behind laser operation, it's time to look at important laser characteristics. Later on, we will look at individual types of lasers.

Quiz for Chapter 3

1. What is the major advantage of a four-level laser over a three-level laser?
   a. More levels to excite atoms to
   b. The lower laser level is not the ground state
   c. More metastable states
   d. No advantage

2. If you put one watt of laser power into a 10-centimeter long oscillator with gain of 0.1 per centimeter, what will be the output power, neglecting losses and saturation?
   a. 1 watt
   b. 1.1 watts
   c. 10 watts
   d. 11 watts
   e. 12.6 watts

3. What is the emission angle of a laser rod (without mirrors) that is 1 centimeter in diameter, one meter long, and has a refractive index of 1.5?
   a. 0.005 degree
   b. 0.43 degree
   c. 1.02 degrees
   d. 1.5 degrees
   e. None of the above

4. The round-trip loss of helium-neon laser light in a cavity is 2%. The output mirror lets 1% of the light escape in the beam. When the laser is operating in a steady state, what is the round-trip amplification in the laser, measured as a percentage?
   a. 1%
   b. 2%
   c. 3%
   d. 5%
   e. 99%

5. A helium-neon laser has intracavity power of 100 milliwatts and output power of 1.5 milliwatts. Neglecting cavity losses, what is the fraction of light reflected back into the cavity by the output mirror?
   a. 1.5%
   b. 50%
   c. 90%
   d. 98.5%
   e. None of the above

6. The helium-cadmium laser has a wavelength of 442 nm. What is the round-trip length of a 30-cm long cavity measured in wavelengths?
   a. 13
   b. 300
   c. 475,000
   d. 1,000,000
   e. 1,357,000

7. Which of the following resonator types is not a stable resonator?
   a. Plane-parallel
   b. Concentric
   c. Confocal
   d. Hemispherical
   e. Hemiconfocal

8. How many internal minimum-intensity points are there in a TEM_{03} mode beam?
   a. None
   b. 1
   c. 2
   d. 3
   e. 6

9. What type of laser cavity could produce a beam with a central dark spot?
   a. Plane-parallel
   b. Confocal
   c. Unstable
   d. Stable
   e. Concentric

10. Which of the following factors does not harm laser efficiency?
   a. Atmospheric absorption
   b. Excitation energy not absorbed
   c. Problems in depopulating the lower laser level
   d. Inefficiency in populating upper laser level

Laser Characteristics

ABOUT THIS CHAPTER

Now that we have seen what goes on inside lasers, it's time to look at what you see when you use a laser. In this chapter, we will learn about important properties of laser light, including coherence, wavelength, directionality, beam divergence, power, modulation, and polarization. We also will learn about the factors that determine laser efficiency.

COHERENCE

Coherence probably is the best-known property of laser light. Light waves are coherent if they are in phase with each other—i.e., if their peaks and valleys are lined up at the same point, as shown in *Figure 4-1*. Two things are necessary for light waves to be coherent. First, the light waves must start out having the same phase at the same position. Second, their wavelengths must be the same, or they will drift out of phase because the peaks of the shorter wave will arrive slightly ahead of the peaks of the longer wave.

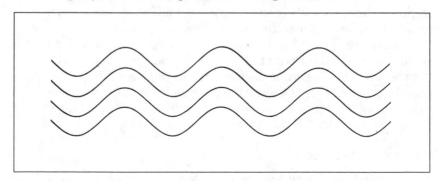

Figure 4-1. Coherent light waves.

As we learned earlier, laser light is coherent because stimulated emission is coherent with the light wave or photon that stimulates it. The stimulated wave has the same phase and wavelength. It, in turn, can stimulate the emission of other photons, which are in phase and have the same wavelength both with it and the original wave.

Coherence of Laser Light

The preceding statement might make it sound like all laser light is perfectly in phase, but this is not the case. Not all photons in a laser beam are descended from the same original photon by stimulated emission, so they all don't start out in phase. Even if they were, the uncertainty principle leads to tiny variations in the wavelengths of the emitted light, which accumulate and become significant after the light goes a long distance. In addition, tiny fluctuations within the laser—such as thermal gradients or vibrations—can affect one light wave in a different way than others, degrading coherence. Thus, there is no such thing as "perfect" coherence.

Not all laser light is equally coherent. In fact, the light from some lasers is almost incoherent. The reasons lie inside the lasers themselves. Some lasers emit a broader range of wavelengths than others, and as we saw earlier, light waves must have the same wavelength to be coherent. Thus a laser should oscillate in only one longitudinal mode to produce the most coherent output because the slightly different wavelengths of other longitudinal modes reduce the degree of coherence. Lasers with high gain tend to emit a broad range of wavelengths because emission can be stimulated by a range of wavelengths (generated by spontaneous emission). In other words, high-gain lasers are likely to amplify many different spontaneously emitted light waves. Because of the uncertainty principle, the range of wavelengths in a pulse increases as the length of a pulse decreases, so the shortest pulses tend to have the broadest wavelength ranges. On the other hand, continuous-wave beams can have the narrowest wavelength range. Add this all together and you find that the most coherent beams come from continuous-wave, low-gain lasers operating in the lowest-order TEM_{00} mode.

Types of Coherence

If you look closely, there actually are two kinds of coherence: temporal and spatial. "Temporal" coherence measures how long light waves remain in phase as they travel (the term "temporal" comes because the degree of coherence is compared at different times). Light waves become incoherent as differences in their optical paths or wavelengths make them drift out of phase. All light has some temporal coherence, but only over a characteristic "coherence length," which is very close to zero for ordinary light bulbs but can be many meters for lasers. The coherence length depends on the light's nominal wavelength (λ) and on the range of wavelengths ($\Delta\lambda$) emitted:

$$\text{Coherence Length} = (\text{Wavelength})^2 / 2 \times \text{Wavelength Range}$$

or

$$\text{Coherence Length} = \lambda^2 / 2 \, \Delta\lambda$$

We can convert this equation to calculate coherence length from frequency units, because a laser's "bandwidth" often is given in frequency terms:

$$\text{Coherence Length} = c / 2 \, \Delta\nu$$

where,
   $\Delta\nu$ is the range of frequencies,
   c is the speed of light.

We can see what this means for a few representative sources. Light bulbs emit light from the visible well into the infrared. If we assume the range of wavelengths is 400 to 1000 nanometers and take an average wavelength of 700 nm, we find that a light bulb has a suitably short coherence length of 400 nm. An inexpensive semiconductor laser might have a wavelength of 800 nm and wavelength range of 1 nm, giving a coherence length of 0.3 mm. An ordinary helium-neon laser has a much narrower line width of about 0.002 nm at its 632.8-nm wavelength, corresponding to a coherence length of 10 centimeters. Stabilizing a helium-neon laser so it emits in a single longitudinal mode limits the line width to about 0.000002 nm, and thus extends the coherence length to about 100 meters.

   Spatial coherence, on the other hand, measures the area over which light is coherent. Strictly speaking, it is independent of temporal coherence. If a laser emits a single transverse mode, its emission is spatially coherent across the diameter of the beam, at least over reasonable propagation distances.

Coherence and Interference

Light waves must be somewhat coherent for us to detect the interference effects described in Chapter 2. If you superimpose many incoherent light waves of different wavelengths, the interference effects average out, and they add together to form white light. This is why we don't see interference effects in everyday things, like scenes illuminated by light bulbs or the sun. However, if the light waves are coherent enough to maintain the same phase relative to each other (even though they are not all lined up with peaks and troughs at the same points), adding them together does produce interference effects. At some points, the total amplitudes cancel, producing a dark zone; at other points, the amplitudes add together to produce a bright zone.

These interference effects are desirable in many measurement applications, because they let us measure distance by counting in units of wavelength. Because light waves are so small, this lets us measure distance very precisely. The best-known application that requires coherent light is holography, which is described briefly in Chapter 10.

Coherence is not always desirable, however. When coherent light waves pass through turbulent air, or are reflected from many types of surfaces, they form grainy patterns called "speckle." This shifting speckle is an interference pattern created by slight differences in the paths that light rays travel. It contains information on the quality of the air and the surface, but for most practical purposes it is merely background noise. Such coherent effects make illumination by coherent light inherently uneven, and thus undesirable for many applications.

LASER WAVELENGTHS

Laser light usually is considered "monochromatic," meaning single-colored, but in practice laser light is not perfectly monochromatic. Some lasers can emit on two or more transitions, and as we saw in Chapter 3, stimulated emission generates a range of wavelengths on each transition. The range of wavelengths (or "bandwidth") depends on the nature of the transition. Oscillations within a laser cavity further narrow the range of wavelengths emitted on each transition, although laser cavities may allow oscillations at two or more distinct wavelengths. Thus lasers normally emit a range of wavelengths, which can be narrow or broad, and which can be changed (either made narrower or moved) by adjusting the laser's optics. Nonetheless, the bandwidth of even the broadest-band laser emission is much narrower than that of ordinary light sources such as the sun.

To understand the nature of laser bandwidth, we need to take a closer look at some things that happen inside lasers.

Transitions and Gain Bandwidth

Our earlier discussions of laser transitions may have made it seem that they were abrupt spikes that occurred only at one specific wavelength. In reality, a laser transition is not an abrupt spike, but a curve with a pronounced peak, such as shown in *Figure 4-2*. In *Figure 4-2* two closely related characteristics are shown. The broader curve is the net gain on the transition; that is, the amount of amplification possible in the laser medium as a function of wavelength. The narrower one is the intensity of the laser output as a function of wavelength. (The peak of the laser emission curve is reduced to fit the same scale.)

The much sharper peak of the emission curve is a consequence of amplification and oscillation within a laser cavity. The difference between the gain and emission curves depends on the cavity design. As light passes through the laser medium, stimulated emission is most likely at the wavelengths where gain is highest. Suppose, for example, the gain was 0.1 per centimeter at 500 nm and 0.01 per centimeter at 505 nm. After travelling through 100 cm of the laser medium, that 10-to-1 gain ratio alone would produce a 5000-to-1 difference in output power, if the initial powers were equal (which would not be the case because initial power would be lower where gain was weaker). That effect makes the emission curve much sharper than the gain curve.

If the laser cavity optics do not select any particular wavelength, laser intensity will be highest at the wavelength where the gain is the highest. However, laser action can occur at any wavelength where gain is enough larger than zero to overcome losses in the laser cavity. Thus, the cavity can select a wavelength away

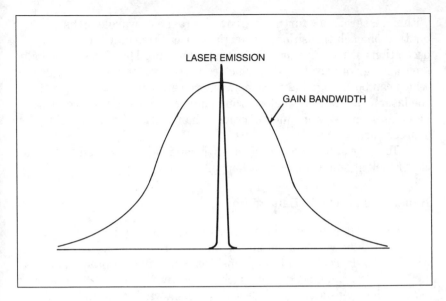

Figure 4-2. Gain bandwidth and laser emission on a laser transition.

from the peak of the gain curve. As we will see later, this allows tuning or adjustment of the output wavelength.

There is one important qualification to this picture of how laser gain makes the intensity of the emitted light vary sharply with wavelength. The effect is not as strong for transitions between different pairs of energy levels. If two transitions have the same upper energy level, stimulated emission is likely to occur first on the stronger one, depopulating that level and making stimulated emission less likely on the weaker level. However, if they have different upper levels, the upper laser level of the weaker transition will not be affected by stimulated emission on the stronger transition. This allows simultaneous emission on different transitions with different strengths.

Line Broadening

Several effects contribute to the width of a laser's gain curve, and they play different roles in different types of lasers. Some broadening comes from the uncertainties of the quantum world. On a quantum-mechanical level, the likelihood of a transition is a probability function, which is largest at the nominal transition wavelength, and which drops off sharply—but not precisely to zero—as wavelength changes.

Interaction of the laser species with other atoms and molecules causes other broadening. In gases, the degree of broadening increases with gas pressure, because the interval between collisions with other atoms or molecules decreases as pressure rises. These collisions affect the interaction between the light emitter and the photon that stimulates emission, and can alter the amount of energy involved. Most gas lasers normally operate at low pressures, but in some high-pressure gas lasers, this effect can make laser emission lines merge into a continuous spectrum. In a solid laser, the transfer of vibrational energy to other atoms can alter emission wavelength.

Another line broadening mechanism is the inherent motion of atoms and molecules, which can shift wavelength by a process called the Doppler effect. Suppose an atom moves toward you while emitting at a constant frequency ν, which normally corresponds to a wavelength of 500 nanometers. If it moves toward you 1 nm in the interval between successive wave peaks, the next peak it emits is only 499 nm behind the first. You thus see a shorter wavelength (or equivalently, a higher frequency), sometimes called a "blue shift." If the atom is moving away, the emitted wavelength becomes longer (a "red shift" because it shifts the light toward the red end of the spectrum).

Atoms and molecules in a gas are constantly moving randomly, with an average velocity that depends on temperature and mass. This random motion means that emitted light experiences random Doppler shifts, spreading out the range of emitted wavelengths. For a temperature T and a mass M, the Doppler bandwidth, or broadening caused by this effect measured in frequency units, $\Delta\nu/\nu$, is given by:

$$\Delta\nu \ / \ \nu = [(8 \ln 2) \, kT \ / \ Mc^2]^{1/2}$$

where,
  $\Delta\nu$ is the change in frequency,
  ν is the frequency,
  k is the Boltzmann constant,
  c is the speed of light.

This equals a few parts per million for typical gas atoms or molecules. The fractional broadening is the same when measured in units of wavelength.

4

Wavelength Selection in a Cavity

We also saw in Chapter 3 that a simple plane-parallel mirror cavity of length L could support oscillation at many different wavelengths λ as long as they satisfied the equation

$$2L = N\lambda$$

where,
   N is any integer.

As long as the laser cavity is much longer than the wavelength (which is the usual case), it's virtually certain that some wavelength given by the equation will fall within the active medium's gain curve. In fact, as shown in *Figure 4-3*, two or more oscillation wavelengths usually fall within this curve. Each of these wavelengths is a distinct longitudinal, and each mode has a bandwidth much narrower than the overall laser emission.

We should note we define wavelength λ as wavelength in the laser cavity. This is not the laser wavelength in vacuum because there's something inside the laser cavity—the laser medium, which has a refractive index n. As we saw earlier, the wavelength of light in a material with refractive index n, λ_n is equal to the wavelength in a vacuum (λ) divided by the refractive index, or λ/n. Thus to be more accurate, the equation defining possible wavelengths in a laser cavity should read:

$$2nL = N\lambda$$

where,
   n is the refractive index in the cavity,
   L is the cavity length,
   λ is the wavelength (in vacuum),
   N is an integer.

This equation is precise only if the laser medium is a single material with uniform refractive index n, but it does indicate the role of refractive index in determining wavelength. As we saw earlier, for a gas the refractive index is very close to one, but it is close to 1.5 in glass, and is even higher in some other transparent materials.

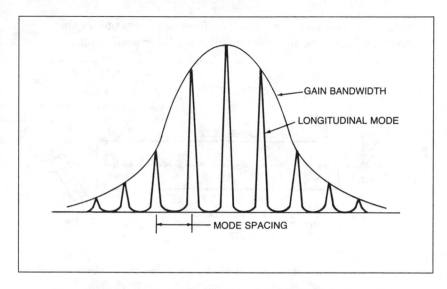

Figure 4-3. Multiple longitudinal modes fall within the gain bandwidth of a gas laser.

The actual profile of wavelengths emitted by a laser is the product of the envelope of longitudinal oscillation modes and the gain profile. The wavelengths to the sides of the gain curve tend to be weaker than those in the center, for the reasons we described before. The profile in *Figure 4-3* is based on typical emission from a gas laser, where a few longitudinal modes can oscillate. In some other lasers a single central longitudinal mode will dominate, with much lower power emitted at one or two peripheral modes.

Single Longitudinal Mode Operation

There are ways to limit laser oscillation to the width of a single longitudinal mode, which is about one megahertz or (for helium-neon lasers) 0.000001 nanometer (10^{-15} meter). The most common technique is to insert a pair of parallel reflective surfaces between the laser mirrors to form a secondary resonant cavity. These added surfaces create what is called a Fabry-Perot etalon; they are tilted at an angle to the axis of the laser medium, as shown in *Figure 4-4*. At most wavelengths, the etalon reflects some light away from the cavity mirrors, increasing cavity loss and suppressing laser oscillation.

However, at certain wavelengths interference effects nullify the reflection, eliminating the loss and allowing laser oscillation.

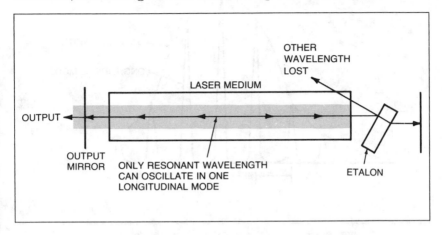

Figure 4-4. A Fabry-Perot etalon limits a laser cavity to oscillation in a single longitudinal mode.

Etalons are designed to allow laser oscillation at wavelengths far enough apart that only one will fall under the laser's gain curve. This limits laser oscillation to a single longitudinal mode. The wavelength of the low-loss mode can be tuned by adjusting the spacing between the Fabry-Perot reflectors, or by turning the device in the laser cavity. (Many Fabry-Perot etalons are solid blocks with reflective surfaces, which can be tuned only by turning them.)

Laser Wavelength Stability

Laser wavelengths are not absolutely stable. The transition wavelengths are constant, but as we have seen, other factors influence the actual oscillation wavelength. For example, temperature fluctuations may cause thermal expansion and slightly change the length of a laser cavity. Temperature changes also can change the effective length by changing the laser medium's refractive index.

The gradual drift of laser wavelength can change how much light is emitted at the wavelengths corresponding to different longitudinal modes. In some cases, this can lead to "mode hopping," where the laser shifts quickly from one dominant mode to another, at a slightly different wavelength.

Single and Multiwavelength Operation

We have seen that the nominal "single wavelength" emitted by a laser is not really a single wavelength but a group of wavelengths in a narrow band defined by a single transition. In some cases, a family of closely spaced transitions can produce emission at a number of closely spaced wavelengths, or allow the wavelength to be tuned continuously through a range or "band." In other cases, the emission is at several discrete wavelengths, which may involve different upper and/or lower laser levels. The results differ from the user's standpoint.

The two most important examples of how transitions can overlap to form a continuous tuning range are the tunable dye laser and the high-pressure carbon-dioxide laser. In later chapters, we will describe them in more detail, but here we will concentrate on their output wavelengths.

In both lasers, the important laser energy levels are broken up into many sublevels. The carbon-dioxide laser is the more straightforward. The main transition occurs when the carbon-dioxide molecule shifts from one vibrational mode to another. (There are actually two such transitions close to one another, one centered near 10.5 micrometers, the other near 9.4 micrometers.) While shifting vibrational states, the molecule also changes rotational energy levels. The rotational levels are much smaller than the vibrational transitions, so at low pressures an untuned carbon-dioxide laser can emit many wavelengths, as shown in *Figure 4-5*. Each wavelength is produced by a transition between different rotational sublevels during a vibrational transition.

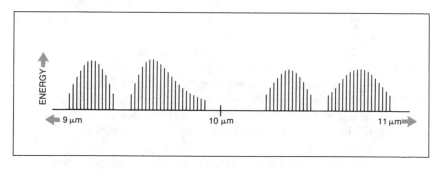

Figure 4-5. The many wavelengths emitted by a carbon-dioxide laser near 10 micrometers.

As we mentioned earlier, if you increase laser gas pressure, the transitions will spread over a wider range of wavelengths. In carbon dioxide, the broadening is large enough and the wavelengths of the transitions close enough that eventually they overlap to form a continuous range of wavelengths.

Even the natural line-narrowing we described earlier does not prevent these lasers from emitting at multiple wavelengths. It is possible to adjust the cavity optics so they only allow oscillation at some of these wavelengths, but that comes at the cost of lower output power, because it limits the number of atoms or molecules from which stimulated emission can extract energy.

Many lasers can emit light on two or more transitions and widely separated wavelengths. In some, the transitions are between entirely different pairs of energy levels. In others, the transitions may involve some common energy levels. For example, neon atoms in one metastable state in the helium-neon laser can produce stimulated emission on transitions to lower states at wavelengths of 543, 632.8, and 3391 nanometers. The laser cavity optics determine the transition on which such a laser will oscillate. Note that it is possible for some such lasers to oscillate on two or more transitions at once.

Laser Wavelength Tuning

At the end of Chapter 3, we mentioned how lasers could be tuned to emit different wavelengths. The basic technique is simple but powerful, and deserves a bit more explanation now that we have talked about the concept of gain bandwidth.

Laser cavity optics normally reflect light over a bandwidth that is wide compared to the laser's gain bandwidth. Inserting a prism or diffraction grating into the laser cavity effectively narrows the cavity's oscillation bandwidth. The prism or grating bends light of different wavelengths at different angles. As was shown in *Figure 3-13*, only one narrow range of wavelengths is aligned properly to oscillate back and forth within the laser cavity (including the prism or grating). In effect, the mirror or prism increases losses at all other wavelengths by deflecting light out of the laser cavity, thus preventing oscillation.

Turning the grating or prism, or some other optical element in the laser cavity, as we saw earlier, changes the wavelength that can oscillate in the laser cavity. As long as that wavelength remains in the laser's gain bandwidth, the laser can operate. However, if the wavelength falls outside the gain bandwidth, the laser cannot oscillate. Because all laser media have gain only over limited ranges of wavelengths, their tuning ranges are limited. For example, one dye might allow tuning of a dye laser only in the orange part of the spectrum. To get different colors, you would have to substitute another dye.

As we saw earlier in this chapter, gain varies with wavelength. If you tune the wavelength of a laser, the gain within the laser medium will change. Because gain influences the power level from the laser, this means that output power also will change as the laser's wavelength is changed.

LASER BEAMS AND MODES

Another important characteristic of lasers is that their emission is concentrated in a narrow beam. As we learned in Chapter 3, the nature of this beam depends largely on the laser cavity. Now let's look more closely at the characteristics of laser beams outside the laser cavity.

Beam Divergence, Diameter, and Spot Size

We think of laser beams as tightly focused and straight, but once they go far enough from the laser, they actually spread out slightly with distance. The spreading angle is called the beam divergence. It usually is measured not in the familiar unit of degrees, but in radians, or more properly milliradians (thousandths of a radian). The radian is a unit derived from circular measurements. A full circle equals 2π radians, and one radian equals $57.2958°$. One milliradian is $0.0572958°$, or 3 arcminutes 33.36 arcseconds. Radians are convenient because the sine of a small angle roughly equals its measure in radians. The divergence of most continuous-wave gas lasers is around a milliradian, but it usually is larger for pulsed lasers. (For the few lasers where the beam divergence is much larger, it often is written in degrees.)

Near Field Conditions

The diffraction-caused beam divergence that we mentioned in the last chapter, and will describe more later, occurs only in what is called the "far field," at comparatively large distances from the laser.

In the near field, the beam shows little spreading, and remains essentially a bundle of parallel light rays.

The distance from the laser over which the light rays remain parallel is sometimes called the *Rayleigh range*. It depends roughly on the beam diameter d and the wavelength λ:

$$\text{Rayleigh Range} = d^2 / \lambda$$

For a visible beam with nominal wavelength of 500 nm, this range is 2 meters if the beam is one millimeter in diameter, and 50 meters if the beam diameter is 5 mm.

Far-Field Beam Divergence

Beyond the Rayleigh range, in the laser's far field, beam divergence becomes the critical parameter in calculating beam diameter, as shown in *Figure 4-6*. Divergence angle normally is measured from the center of the beam to the edge. The edge itself must be defined because beam intensity gradually drops off at the sides. The usual definition of the beam's edge is the point where intensity drops to $1/e^2$ of the maximum value.

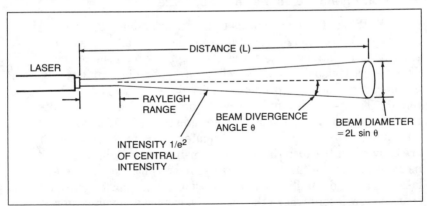

Figure 4-6. Divergence of a laser beam, exaggerated to make it visible.

Calculating the size of the beam is a matter of trigonometry. Simply multiplying distance by the sine of the divergence angle gives the beam radius, so a factor of two must be added to give diameter:

$$\text{Diameter} = 2 \times \text{Distance} \times \sin(\text{Divergence})$$

or, in more properly mathematical form

$$D = 2\,L\,\sin\theta$$

where,
D is the beam diameter,
L is the distance from the laser to the spot,
θ is the divergence angle.

If the divergence is under about 0.1 radian (6 degrees), you don't have to bother calculating the sine; the angle in radians is a good enough approximation. That is the case for most lasers except semiconductor types.

Beam Waist and Divergence

There are several ways to calculate beam divergence. If the resonator mirrors are curved, the beam has a narrow point or "waist" inside the laser cavity. At large distances from such lasers, we can approximate the divergence angle using the formula

$$\text{Divergence} = \text{Wavelength} / \pi \times \text{Waist Diameter}$$

or

$$\theta = \lambda / \pi W$$

where,
θ is beam divergence,
λ is wavelength,
W is diameter of the beam waist.

The beam waist diameter itself depends on laser wavelength, cavity length, and design. For a simple confocal resonator (with two mirrors both having their focal points at the midpoint of the cavity) of length L, the beam waist diameter W is

$$W = (L\lambda / 2\pi)^{0.5}$$

or about 0.17 millimeter for a 30-centimeter (one-foot) long helium-neon laser emitting at 632.8 nanometers. This relation holds for lasers with resonant cavities that produce good-quality beams. The best beams are said to be "diffraction limited," because their minimum divergence is limited by diffraction effects.

Output Port Diameter and Beam Divergence

Although it is helpful to understand these relationships, in practice you will be given a laser's beam diameter and divergence. The diameter is measured at the laser's output port. It can be changed by optical systems that expand or contract the laser beam but still leave the light all going in the same direction. The divergence depends on how intensity varies across the beam as well as on the diameter at the output port, but a good approximation for diffraction-limited beams is:

$$\text{Divergence} = \text{Wavelength} / \text{Diameter}$$

or

$$\theta = \lambda / D$$

where,
D is the diameter of the output optics.

Remember from the last chapter that a more precise formula is

$$\theta = K\lambda / D$$

where,
K is a constant somewhere in the range of 1 that depends on the variation of intensity across the beam.

Note the implications of this formula for beam divergence. The larger the output optics, the smaller the beam divergence at long distances. Thus you need large optics to focus light onto a small spot. Conversely, the shorter the wavelength, the smaller the spot size. In practice, this means that you can't have both tiny spot diameter and narrow divergence. If you start with a one-millimeter diameter beam from a helium-neon laser, for example, its minimum divergence is about 0.6 milliradian.

Divergence can be larger than these formulas indicate if the beam contains multiple transverse modes, which tend to spread out more rapidly than a single-mode TEM_{00} beam. Remember that lasers without resonant cavities have beams that diverge more rapidly.

By the way, the reason we're being vague about the value of the constant K is because it changes with the distribution of light across the emitting area or laser beam. If the light is equally bright across the entire opening, its value is 1.22. However, the value is smaller if the intensity drops off smoothly with distance from the center of the beam.

Focusing Laser Beams

Our earlier descriptions of focusing by lenses might have made you think that they could focus a laser beam down to a perfect point of zero diameter. However, even a perfect TEM_{00} laser beam cannot be focused that tightly. The minimum diameter of a focal spot S formed by a lens of diameter D and focal length f with light of wavelength λ is

$$S = f\lambda / D$$

If you're familiar with photography, you may recognize the ratio f/D as the focal ratio or "f number" of a lens. Thus for ordinary lenses, this formula implies that the smallest focal spot is a few times the wavelength of light. (It is hard to make lenses with focal ratios smaller than about 2.)

Transverse Mode Effects

In talking about beam diffraction, we have assumed that the beam is in the TEM_{00} transverse mode, where intensity drops smoothly from a peak at the center. However, as we saw in Chapter 3, there is a whole family of TEM_{mn} transverse modes.

The crucial factor for far-field divergence turns out to be how many transverse modes the laser emits simultaneously. As the number of transverse modes (N) in the beam increases, so does the far-field divergence. The divergence increases roughly by the square

4

root of the number of modes ($N^{\frac{1}{2}}$), and the area of the illuminated spot is thus roughly proportional to N.

Longitudinal Modes

We saw earlier that lasers have longitudinal as well as transverse modes. These modes are actually different wavelengths that lie within the laser's gain bandwidth and satisfy the equation for oscillation in a laser cavity with length L.

$$2L = N\lambda$$

Figure 4-3 showed how the narrow bandwidths of longitudinal modes compared to the broader range of gain bandwidth. Each wavelength spike is called a "longitudinal mode."

Because lasers have both transverse and longitudinal modes, references to "single" and "multimode" operation can be confusing. In practice, those terms usually refer to transverse rather than longitudinal modes. Lasers oscillating in a single longitudinal mode often are called single-wavelength or single-frequency lasers.

Remember that a laser operating in a single transverse mode can oscillate in two or more longitudinal modes. Indeed, this is common. Many gas lasers fall into this category, emitting a TEM_{00} single transverse mode beam that contains two or more longitudinal modes. As we saw earlier in this chapter, you usually need a special element called a Fabry-Perot etalon to restrict laser oscillation to a single frequency.

Polarization

The polarization of light is the alignment of the electric and magnetic fields that make up a light wave. There are several types of polarization, but the simplest to understand—and most relevant to introductory laser physics—is linear polarization. Light waves that are linearly polarized all have their electric fields aligned in the same direction. One peculiar property of light waves is that they always can be separated into components with linear polarizations that are perpendicular to each other (although those polarizations can change with time). Thus unpolarized light can be divided into two components with linear polarization, one with a vertical electric field and one with a horizontal electric field.

Normally, most lasers emit unpolarized light. However, addition of a component that selects between the two perpendicular polarizations lets a laser generate linearly polarized light. The usual choice is a Brewster window, shown in *Figure 4-7*. This is a surface aligned at a critical angle at which different things happen to light waves with different polarization. If the polarization is in the plane of incidence, (the plane of the paper in *Figure 4-7*), no light is reflected. However, about 15% of the light is reflected if the plane of polarization is perpendicular to the angle of incidence on a glass surface.

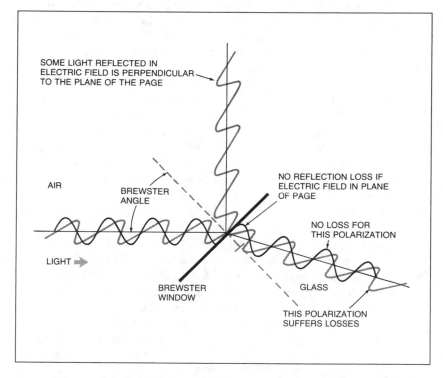

SOME LIGHT REFLECTED IN ELECTRIC FIELD IS PERPENDICULAR TO THE PLANE OF THE PAGE

AIR

BREWSTER ANGLE

NO REFLECTION LOSS IF ELECTRIC FIELD IN PLANE OF PAGE

NO LOSS FOR THIS POLARIZATION

LIGHT ➡

BREWSTER WINDOW

GLASS

THIS POLARIZATION SUFFERS LOSSES

Figure 4-7. Operation of a Brewster window.

A Brewster-angle window makes a critical difference in the operation of a low-gain laser if it is placed at the end of the tube. The Brewster window makes gain different for light of different polarizations, so the laser emits light polarized in the direction of lower loss. The 15% loss from the window eliminates one linear polarization from the output, and leaves the other. Note that designing the laser to produce a linearly polarized output beam does not reduce the output power—it lets the laser operate with lower internal loss, and thus may let it generate more power than might otherwise be possible.

OSCILLATORS AND AMPLIFIERS

So far we have assumed that all lasers are oscillators, with a pair of cavity mirrors, one totally reflective, and an output mirror that transmits a fraction of the incident light which emerges as the laser beam. As we mentioned earlier, you can have a population inversion and stimulated emission without mirrors. In fact, lasers sometimes are used as amplifiers.

The basic arrangement is shown in *Figure 4-8*. The laser on the left is an oscillator, with a totally reflective rear mirror and a partly transparent output mirror. The beam from the oscillator passes through another cavity containing the same laser medium, but without mirrors. In our drawing, the oscillator beam makes a single pass through the amplifier, stimulating emission from atoms or molecules in the upper laser level. Such an oscillator-amplifier arrangement can boost the power available from lasers with high enough gain that a single pass through the laser medium will amplify the light significantly.

One functional difference between oscillators and amplifiers is that oscillators can generate their own beams, while amplifiers cannot. However, as we learned earlier, you do not need a highly reflective pair of mirrors if the laser has high gain. In fact, some types of lasers work well with a totally reflective rear cavity mirror and an output window that reflects only a small fraction of the light back into the laser cavity and transmits the rest in the laser beam.

OUTPUT POWER

From the standpoint of laser users, the power level in the beam is one of the most important laser characteristics. This is a more complex matter than it may seem at first, because it depends upon several things, including variations in power level with time, efficiency of converting excitation energy into laser energy, excitation methods, and size of the laser itself. If we look at each of these

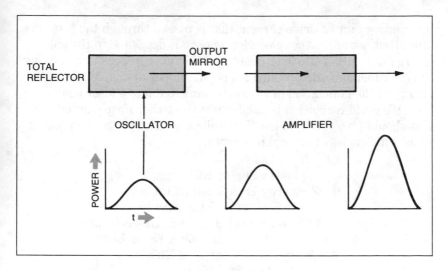

Figure 4-8 Combination of a laser oscillator with an amplifier.

factors individually, we also find that we may have to look a bit deeper.

Laser Efficiency

Efficiency is the sort of word that you think you understand until you have to measure it. That is especially true with lasers. The most useful practical measurement is often called "wall-plug" efficency, and measures how much of the energy put into the laser system (through the wall plug) comes out in the laser beam. Unfortunately, this is not always the efficiency that people measure— even when they use the term "overall" efficiency. Sometimes other types of efficiency are given, including "slope" efficiency, "excitation" efficiency, and "conversion" efficiency. All those factors enter into the equation for wall-plug efficiency.

Wall-Plug Efficiency

Let's start by looking at all the factors that enter into wall-plug efficiency. For simplicity, we'll consider a gas laser, but the same principles work for other lasers as well. Electrical power enters the laser through a power supply, which converts much (but not all) of the input energy into a drive current that is passed through the laser gas. Much (but not all) of that electrical energy is deposited in the gas (energy-deposition efficiency). Much of that energy goes into exciting the laser medium and producing a population inversion. Much of the energy in the population inversion is converted into a laser beam.

We could carry that breakdown even further if we wanted, but that should give you the idea. The wall-plug efficiency is the product of the efficiencies of several processes:

$$\begin{aligned}
&\text{Power Supply Efficiency} \\
\times\ &\text{Energy-Deposition Efficiency} \\
\times\ &\text{Excitation Efficiency} \\
\times\ &\text{Fraction of Energy Available From Laser Transition} \\
\times\ &\text{Fraction of Atoms That Emit Laser Light} \\
\times\ &\text{Cavity Output-Coupling Efficiency} \\
=\ &\text{Overall Efficiency}
\end{aligned}$$

Plug in some numbers and you can see why lasers have low wall-plug efficiencies. If the power supply, energy deposition, and excitation are each 80% efficient, the laser transition represents 50% of the excitation energy, half the atoms emit laser light, and 80% of the laser energy is coupled out of the cavity in the beam, (all fairly optimistic assumptions), the wall-plug efficiency is $0.8 \times 0.8 \times 0.8 \times 0.5 \times 0.5 \times 0.8 = 0.1024$, or 10%. Let's look briefly at the major elements in the efficiency equation.

Power-Supply Efficiency

No power supply is 100% efficient. In general, the more manipulation of the input electrical energy, the lower the efficiency. Conversion of 120-volt alternating-current input to the low-voltage direct current needed to drive a semiconductor laser is straightforward. Likewise, conversion to steady high voltages to drive a continuous-wave gas laser is efficient. However, losses can be significant if the power supply must generate intense electrical pulses for an electron-beam driven pulsed laser.

The biggest source of losses in the "power supply" category is converting electricity into other forms of energy. For example, solid-state crystalline and glass lasers require optical excitation. This light is generated by passing electricity through a flashlamp or arc lamp (or even using electricity to power another laser). This conversion process normally is inefficient, and contributes important losses.

Excitation Efficiency

Some energy that enters the laser medium does not excite atoms or molecules to the upper laser level. One reason is that some of the light or electricity passes through the laser medium unabsorbed. Some absorbed energy does not put atoms or molecules into the right state to emit on the laser transition; for example, it might heat a gas rather than raise it to the upper laser level.

This absorption efficiency depends on the laser design. The tradeoffs can be subtle. For example, it might seem a good idea to have a lot of laser material, to absorb all the energy entering it. However, that would require an excess of light-absorbing material in some parts of the laser cavity, and that excess material would reduce the population inversion needed for laser action. As such tradeoffs indicate, there usually is an optimum value for absorption efficiency that depends on the properties of the laser material and the way it is excited.

Another limitation on excitation efficiency comes from the fact that not all the energy can be absorbed efficiently. Remember in our simple examples of three- and four-level lasers that there was a single pump transition. From a very simplified point of view, the excitation energy must match that transition to excite the laser. Actual requirements aren't that stringent, but the point is that the excitation energy has to be packaged in the right way to be absorbed. For example, a flashlamp emits light across much of the visible spectrum (and parts of the ultraviolet and infrared), but solid-state crystalline lasers can absorb only some of those wavelengths. (At least in ways that produce a population inversion.) Some of this energy may heat the laser medium; some of it may not be absorbed at all.

Laser Transition Efficiency

When we described four-level laser transitions, we showed that laser pumping raised an atom or molecule to an excited level, which dropped down into the upper laser level. After laser emission, the atom or molecule again dropped from the lower laser level to the ground state. If you go back and look carefully at *Figure 3-2*, you'll note that some energy is lost in the process.

To make the point clearer, let's look at a hypothetical atom in

Figure 4-9 and assign each energy level a value. (For simplicity, we'll use energy units, electronvolts.) Suppose the atom initially absorbs 4 eV to raise it to a short-lived upper level, which then drops to the upper laser level at 3 eV. The laser transition takes the atom to the lower laser level at 2 eV, from which the atom drops to the ground state at 0 eV. It takes 4 eV to produce a 1 eV laser photon—meaning that the transition is at most 25% efficient. Some real laser transitions are better than our example, but many are much worse.

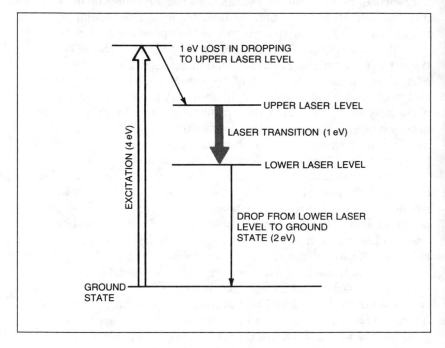

Figure 4-9. How 4 eV of excitation produces only 1 eV of laser energy.

The need to produce a population inversion makes matters even worse, especially in a three-level laser, where the ground state is the lower laser level. For a three-level laser to work, more atoms must be in the upper laser level then in the ground state. Once half the atoms have dropped down to the ground state, it won't emit any more laser light, and the energy left in the remaining atoms in the upper laser level is lost.

Matters are not as bad in four-level lasers where ways exist to get atoms and molecules out of the lower laser level. However, laser action cannot occur in any system until a population inversion is produced. This requires depositing enough energy to pass what is called a threshold for laser action. Once the threshold is passed, some of the input energy comes out as laser light, but below the laser threshold, no laser light is produced.

We show how laser threshold works by plotting input energy vs laser output in *Figure 4-10*. The threshold phenomenon strongly affects laser efficiency. When input power is just above threshold, little output power emerges, and the laser is quite inefficient. However, adding extra input power raises the output power, by an increment called the "slope efficiency." For example, if you need ten

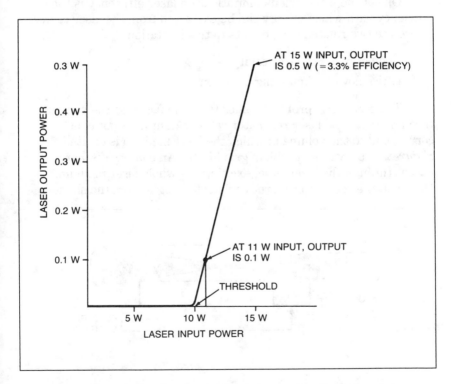

Figure 4-10. Laser output does not emerge until the input energy has passed a threshold at 10 W. Then it rises with 10% slope efficiency.

watts of input power to pass the threshold, that much input may produce only 0.001 W of output, for 0.01% efficiency. However, the slope efficiency may be 10%, so adding another watt of input power adds 0.1 W of output power, raising total power to 0.101 watt—for 0.9% efficiency.

Slope efficiency is an important concept, because it shows the fraction of energy above laser threshold that emerges as output. Note, however, that you may need to be careful in defining slope efficiency—because it, too, can denote wall-plug efficiency or other types of efficiency.

Energy Extraction Efficiency

One of the less-obvious limitations on laser efficiency is the efficiency with which energy can be extracted from the laser cavity. There are two related components to this limitation:

1. Geometry of the laser cavity
2. Efficiency of stimulating emission

The geometric problem comes from the fact that the volume through which the laser resonator reflects light does not exactly correspond to the volume in which the laser medium is excited. *Figure 4-11* shows the problem graphically. An electric discharge passes through the laser tube, exciting the whole laser medium. However, the resonator mirrors only collect light from the shaded

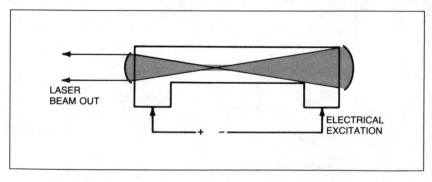

Figure 4-11. Electrical discharge passes through a laser tube, but the resonator only extracts energy from the gas in part of the tube.

area. The excitation energy absorbed outside that area cannot contribute to the laser beam.

It is easy to see why energy cannot be extracted from atoms that lie outside the volume where the resonant cavity mirrors collect energy. The distribution of energy need not be uniform. For example, energy is lowest in the outer portions of a standard Gaussian beam. Inside the cavity, this means that toward the edges of the beam there are fewer photons to stimulate emission than in the center. Assuming that the energy is deposited uniformly, this means that less of the available energy will be extracted toward the edges of the beam. One attraction of unstable resonators is their ability to collect laser energy throughout the laser cavity.

Waste Heat

As we have shown, most energy that goes into a laser does not emerge in the beam. In fact, a good laser efficiency is about 20%, while many lasers are less than 1% efficient, as shown in *Table 4-1*.

Table 4-1. Efficiency Range of Commercial Lasers

Type	Wall-Plug Efficiency
Argon ion	0.001–0.01%
Carbon dioxide	5.0–20.0%
Copper vapor	0.2–0.8%
Excimer (rare-gas halide)	1.5–2.0%
GaAlAs semiconductor	1.0–over 10%
Helium cadmium	0.002–0.014%
Helium neon	0.01–0.1%
Neodymium	0.1–1.0%
Ruby	0.1–1.0%

All this energy does not vanish—most of it turns into heat, which must be removed from the laser. Convection cooling by the air can remove waste heat from low-power lasers, but as power (and waste-heat) levels increase, so do cooling requirements. Some lasers require forced-air cooling with a fan; others require cooling with flowing water or some other refrigerant. (We did not consider the cooling system's energy needs in our rough estimate of efficiency in *Table 4-1*.)

Pulsed vs Continuous Emission

Lasers can emit light steadily, operating in what is called the continuous-wave mode, or they can produce pulses. The pulses may be single or repeated at a steady (repetition) rate. The precise characteristics depend both on the physics of the laser and the engineering of the laser excitation mechanism.

Some lasers can emit only pulses because the lower laser levels fill up during laser operation. The problem is it takes longer to remove atoms or molecules from the lower laser level than to produce stimulated emission. The lower laser level becomes a bottleneck, with population increasing rapidly, eventually ending the population inversion. This sort of self-terminating laser may be ready soon afterwards to produce another pulse, but it cannot emit light steadily.

Other lasers are limited to pulsed operation by the excitation method. For example, the excimer laser is excited by an electrical discharge flowing through a mixture of gases. That discharge is stable only for a very brief interval, well under a microsecond. Breakdown of the discharge terminates the laser pulse. (In this case, excimer kinetics also limit pulse duration.)

Heat dissipation problems limit other lasers to pulsed operation. Ruby is one example. Its laser characteristics degrade with temperature, so ruby can only generate pulses because heat cannot be removed fast enough if the laser action is continuous.

Other lasers can operate in a steady state, generating a continuous beam. This requires a combination of favorable circumstances: a set of energy levels and transitions that allows continuous operation, a stable way to supply energy continuously to the laser medium, and an efficient way to remove waste heat. The higher the power, the more serious the waste-heat problem becomes. Practical ways to remove this excess heat include flowing the laser gas through the tube, forced-air cooling with a fan pushing air by the laser medium, or cooling with a flowing liquid (water or a refrigerant).

In practice, some lasers operate only in pulsed mode because they cannot sustain steady laser emission. Others operate only continuous wave because it takes time to establish the right conditions for laser oscillation. Some lasers can operate either pulsed or continuously, depending on operating conditions.

Excitation Techniques

Excitation techniques often determine whether laser operation is pulsed or continuous. Laser emission usually occurs almost immediately after excitation starts, and stops almost immediately after the excitation ends. The rise and fall times are not instantaneous, but in pulsed lasers they are typically faster than the excitation pulses. Note, however, that it takes time to establish stable operating conditions for some continuous-wave lasers, as well as for pulsed lasers that must be heated, such as the copper-vapor laser.

The pulse repetition rate depends on the excitation mechanism as well as the internal laser physics. Most gas laser media can dissipate heat and otherwise restore normal conditions faster than it takes a power supply to accumulate enough energy to fire another electrical pulse into the gas. Electrical pulse generation speeds also can limit repetition rates of some flashlamp-pumped solid-state lasers. However, other solid-state lasers are limited by the rate at which heat can be removed from the laser rod. One example is ruby, which typically can generate no more than a pulse or two a second. Because they are very small and dissipate very low levels of power, semiconductor lasers can be modulated very fast—at billions of pulses per second—by modulating their electrical drive currents.

Some lasers can operate pulsed or continuous wave under different conditions. For example, the carbon-dioxide laser emits a continuous-wave beam when excited by a steady discharge at low gas pressure. However, it also can produce short pulses if electrical pulses are passed through the gas mixture at much higher pressures near one atmosphere).

Optical and electrical excitation both allow considerable control over pulse characteristics. Even a nominally continuous-wave laser like the helium-neon laser can be made to emit pulses by turning its power supply off and on. However, particularly for electrical excitation, this can come at a cost in reduced stability, and it may be more practical to modulate the beam intensity outside the laser, as described in the following.

Controlling Pulse Characteristics Externally

The characteristics of laser pulses often are controlled externally rather than by modulating laser excitation. Conceptually, the simplest modulation is by blocking the beam at some times, which can be done in various ways, as described in the next chapter.

Pulse characteristics also can be changed by inserting special optical components into the laser cavity. The most important of these laser accessories are modelockers, Q switches, and cavity dumpers, which are described in Chapter 5.

Power and Energy vs Time

We can think of the difference between pulsed and continuous-wave lasers as a difference in how their output power varies as a function of time. Before we go on, we should look carefully at these variations.

Figure 4-12 plots the variation of power with time for a generalized laser pulse. The pulse shape is fairly typical, with a Gaussian profile that rises slowly at first, then rapidly to a peak, then

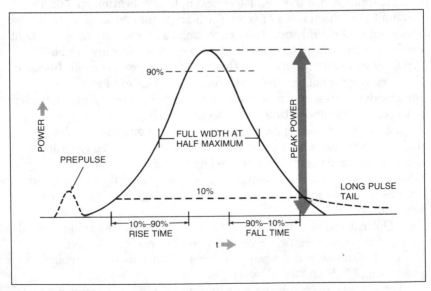

Figure 4-12. Power variation as a function of time (dashed lines show prepulse and long tail).

drops off. A laser pulse has a characteristic rise and fall time. In practice, rise and fall times are measured between the points where laser power is 10% and 90% of the maximum value, not the 0 and 100% points. In some lasers, there may be a smaller prepulse (before the main pulse), and/or a long tail dropping off slowly after the main pulse, which are shown as dashed lines in *Figure 4-12*. The duration of a pulse typically is measured as the "full width at half maximum"—meaning the interval from the time the rising pulse passes 50% of the peak power to the time the power drops below that 50% point.

The peak power is defined as the highest level of optical power in the pulse. It may be attained only for a tiny fraction of the pulse. However, in practice laser specialists often assume that the laser emits its peak power for the full duration of the pulse (as measured at half maximum). As we will see, this greatly simplifies calculations of pulse power and energy.

The energy in a laser beam is the power emitted in a given time. If the power is at a constant level,

$$E = Pt$$

where,
    E is energy (in joules),
    P is power (in watts),
    t is time (in seconds).

Thus, a 0.001-watt laser beam delivers 0.001 joule in one second, 0.002 joule in two seconds, and so on. Because the power generally is a function of time, $P(t)$, the actual energy level is described by an integral

$$E = \int P(t)\, dt$$

taken over the appropriate interval.

For pulsed lasers, it can be useful to know the average level of power the laser emits over an interval in which the beam is both on and off. This can be calculated by adding together pulse energies and dividing by time:

$$\text{Average Power} = \Sigma E / t$$

Alternately, it can be approximated by multiplying peak power by the "duty cycle"—the fraction of the time the laser is emitting light.

$$\text{Average Power} = \text{Peak Power} \times \text{Duty Cycle}$$

Duty cycle in this case is given by:

$$\text{Duty Cycle} = \text{Pulse Length} \times \text{Repetition Rate}$$

Where repetition rate is pulses per second.

To give a concrete example, let's calculate average power of a laser that emits 20 pulses per second, with each pulse lasting 1 microsecond and having peak power of 100 kilowatts. For simplicity, we assume the pulse energy is constant during the pulse. The duty cycle is

$$\begin{aligned}\text{Duty Cycle} &= 10^{-6} \times 20 \\ &= 2 \times 10^{-5}\end{aligned}$$

From that figure, we can calculate average power as

$$\begin{aligned}\text{Average Power} &= (2 \times 10^{-5}) \times (100 \times 10^3 \text{ W}) \\ &= 2 \text{ watts}\end{aligned}$$

We can get the same value by first calculating pulse energy from peak power

$$\begin{aligned}E &= 100 \times 10^3 \text{ watts} / 10^{-6} \text{ second} \\ &= 0.1 \text{ joule}\end{aligned}$$

then multiplying the pulse energy by the repetition rate of 20 pulses per second (usually called hertz) to give

$$\begin{aligned}\text{Average Power} &= 0.1 \text{ joule per pulse} \times 20 \text{ pulses per sec} \\ &= 2 \text{ watts}\end{aligned}$$

Scaling Effects and Power

In Chapter 3, we saw that laser gain was measured as an increment per unit length, for example, 0.02 per centimeter. This means that laser power increases by 0.02 over the original value after passing through one centimeter of laser medium, or it equals 1.02 of the input value. For a medium of length LENGTH, this leads to a formula:

$$\text{Total Power} = (\text{Input Power}) \times (1 + \text{Gain})^{\text{LENGTH}}$$

Remember that this formula is valid only for "small-signal" gain, when the power levels are far from the maximum possible from the medium; output power cannot increase without limit. Actual output power is limited by other factors.

The job of raising laser powers to higher levels is called "scaling," and specialists know that each design can scale over only a limited range. They learned that the hard way, after building laser tubes that (when bends and curves were taken into account) gave gain lengths of up to 750 feet! The designs that work well at low powers often do not work well at much higher powers. Likewise, other designs may make efficient high-power lasers but be of little use at low powers. Some lasers, like helium-neon, work well over only a limited range, from below a milliwatt to about 50 mW. Others, like carbon dioxide, can work from power levels under a watt to tens of kilowatts. However, no one design works for CO_2 lasers over that entire power range. Typically, the lowest-power CO_2 lasers have gas sealed within a tube, but models emitting at tens of kilowatts require rapid gas flow through the laser cavity.

Energy Storage and Gain

The idea of storing energy within a laser cavity can be useful for many types of lasers. Suppose, for example, that the active medium has been excited with a flashlamp, but stimulated emission has yet to start. That laser cavity contains stored energy, in the form of a population inversion, ready to be released as stimulated emission. The amount of stored energy depends on the number of excited atoms or molecules and their excitation energy. It also depends on the excited-state lifetime; the shorter the lifetime, the less energy the laser medium can store.

In practice, the lifetime of the excited state is related to the cross section for stimulated emission, which measures the likelihood of stimulated emission. That cross section, in turn, is related to the

4

laser gain. The higher the cross section, the higher the laser gain. Likewise, the higher the cross section, the shorter the lifetime of the upper laser level. This leads to the somewhat paradoxical result that the higher the gain of the laser medium, the less able it is to store energy.

Pause a moment, however, and you can begin to see why energy storage decreases as gain increases. The higher the gain, the more readily the laser transition occurs—and the harder it is to get energy to stay in the upper laser level.

What Have We Learned?

- Coherent light is in phase and has the same wavelength.

- Temporal coherence indicates how long light waves remain in phase; spatial coherence is the area over which light is coherent.

- Coherence length equals the speed of light divided by twice the bandwidth.

- Light waves must be coherent to display interference effects.

- Coherent noise generates grainy patterns called speckle.

- Although nominally monochromatic, laser light has a nonzero bandwidth.

- The gain bandwidth of lasers is broader than the range of emitted laser wavelengths.

- Thermal motion of atoms and molecules shifts wavelength by the Doppler effect.

- Multiple longitudinal modes fall within the gain bandwidth of a laser transition.

- Wavelength within the laser cavity depends on the refractive index.

- An etalon can restrict a laser to oscillation in a single longitudinal mode.

- Lasers can emit on many overlapping transitions, or on a set of discrete transitions. Many lasers can emit at two or more widely separated wavelengths on different transitions.

- Lasers can be tuned in wavelength with a prism or diffraction grating that lets only one wavelength oscillate in the laser cavity.

- Beam divergence is observed in the far field; in the near field laser beams show little spreading. Far field divergence can be calculated from the diameter of the "waist" inside the laser cavity.

- Beam divergence decreases as diameter of the output optics increases, and increases as wavelength increases.

- Minimum diameter of a focused laser beam depends on laser wavelength, and on focal length and diameter of the lens.

- Beam divergence increases with the number of transverse modes.
- Lasers oscillating in a single longitudinal mode often are called single-wavelength or single-frequency lasers.
- Addition of a Brewster-angle window can linearly polarize laser output.
- Some lasers are used as amplifiers, without resonator mirrors.
- The most useful measure of efficiency compares laser output to overall input energy entering through the wall plug. Wall-plug efficiency is the product of the efficiencies of several processes.
- The need to maintain a population inversion limits laser efficiency.
- Most energy that enters a laser emerges as waste heat.
- Laser pulse characteristics can be controlled by devices inside or outside the laser.
- Laser pulses have a characteristic rise time, pulse duration, and fall time.
- Energy is the integral of power over time. If power is constant, this simplifies to the power level times time.
- Average power is the total energy emitted divided by how long energy is collected.
- Laser output power does not depend on the length of the medium alone. Scaling is complex and differs among lasers.
- Energy can be stored as an inverted population in a laser cavity.

WHAT'S NEXT?

In Chapter 5, we will learn about major laser accessories and how they are used.

Quiz for Chapter 4

1. A semiconductor laser emits 820-nm light with a bandwidth of 2 nm. What is its coherence length?
   a. 822 nm
   b. 0.168 mm
   c. 20 cm
   d. 1 meter
   e. 12.8 meters

2. A single-frequency dye laser has bandwidth of 100 kilohertz (10^5 Hz) and wavelength of 600 nm. What is its coherence length?
   a. 600 nm
   b. 20 cm
   c. 1 meter
   d. 12.8 meters
   e. 1.5 kilometers

3. Which of the following contributes to the broadening of laser emission bandwidth?
   a. Doppler shift of moving atoms and molecules
   b. Amplification within the laser medium
   c. Coherence of the laser light
   d. Optical pumping of the laser transition
   e. None of the above

4. How many longitudinal modes can fall within a laser's gain bandwidth
   a. 1 only
   b. 2
   c. 3
   d. 10
   e. No fixed limit, depends on bandwidth and mode spacing

5. What is the distance over which light from a typical helium-neon laser, with 1-millimeter beam diameter and 632.8-nm wavelength, shows no divergence?
   a. No such distance, beam starts diverging immediately
   b. 0.23 meter
   c. 1.58 meters
   d. 6 meters
   e. 1.5 kilometers

6. A helium-neon laser beam has divergence of one milliradian. What would its diameter be at the 38,000-kilometer altitude of geosynchronous orbit?
   a. 1 meter
   b. 38 meters
   c. 76 meters
   d. 1 kilometer
   e. 76 kilometers

7. Suppose we add a one-meter output mirror to limit the helium-neon laser's beam divergence. What then is the beam diameter at the 38,000-km altitude of geosynchronous orbit? (Assume that $K = 1$)

    a. 1 meter

    b. 48 meters

    c. 94 meters

    d. 1 km

    e. 3.8 km

8. What is the smallest possible spot you can make when focusing a helium-neon laser beam with a lens having 10-cm focal length and 2.5-cm diameter?

    a. 632.8 nm

    b. 1000 nm

    c. 1265.6 nm

    d. 2531 nm

    e. 2 cm

9. What is the maximum wall-plug efficiency for a laser with a power supply that is 70% efficient and energy-deposition efficiency of 25%, in which the laser transition represents 60% of the excitation energy and half the excited atoms emit laser light?

    a. 1%

    b. 5.25%

    c. 10%

    d. 25%

    e. None of the above

10. A pulsed laser generates 500-kilowatt pulses with full width at half maximum of 10 nanoseconds at a repetition rate of 200 hertz. Making simplifying assumptions about pulse power, what is the average power?

    a. 1 watt

    b. 2 watts

    c. 5 watts

    d. 10 watts

    e. 20 watts

Laser Accessories

ABOUT THIS CHAPTER

A variety of accessories helps lasers perform useful functions. In this chapter, we will describe briefly the most important of those accessories. We will start with classical "passive" optics, such as lenses and prisms, then move on to "active" optics which modulate or bend laser light, then finally describe some emerging technologies.

ACTIVE VS PASSIVE OPTICS

It's sometimes useful to distinguish between active and passive optics. Passive optics always do the same thing to light. A lens, for example, always bends light in the same way, and a mirror always reflects light in the same way. Fiber optics, polarizers, filters, telescopes, and prisms are other examples of passive optics.

Active optics can do different things to light at different times. A simple example is a modulator, which changes how much light it transmits depending on the signal that drives it. Another example is a rotating mirror that scans a laser beam across a remote surface. Note the important difference—active optics are controlled actively, while passive optics do not require external control.

We should warn you that some equipment used with lasers—such as light detectors—may not fall neatly into either category. However, the distinction between active and passive optics can be useful.

CLASSICAL PASSIVE OPTICS

At the end of Chapter 2, we learned how lenses and mirrors refract and reflect light. There we talked only about single, simple lenses and mirrors. However, lenses and mirrors can be combined to make much more useful optics. Thorough coverage of the world of classical optics would require a book in itself, but we can give an overview of the field.

Simple Lenses and Mirrors

We usually can consider a laser beam as a set of parallel light rays, although semiconductor lasers are an exception because of their large beam divergence. From the viewpoint of classical optics, parallel light rays are equivalent to light emitted by an object an infinite distance away.

A single positive lens (or a concave mirror) focuses parallel light rays to a small spot at its focal point. This lets you concentrate laser power onto a small spot, with size that depends on the size of the optics and the wavelength of light. The minimum spot size is for a lens with focal length f and diameter D, for a laser wavelength λ is given by the formula we learned in the last chapter:

$$S = f\lambda / D$$

That is comparable in size to the wavelength.

In the same way that a simple lens or mirror can focus parallel light rays to a spot, it also can manipulate light rays going in the opposite direction. Thus, a lens or mirror can bend light from a bright spot or "point source" to form a parallel beam. This is how a flashlight lens or searchlight mirror forms light from a bulb into a narrow beam.

Beam Collimators and Telescopes

Though it may seem amazing today, about two millennia passed between the discovery that lenses or mirrors could focus light and the discovery of the telescope. The ancient Greeks knew about burning glasses, but the telescope was invented about 1600 A.D. It seems doubtful that either discovery was the outcome of a systematic investigation. Legend tells us that children playing with lenses in the shop of Dutch spectacle maker Hans Lippershey discovered the Galilean telescope.

Telescopes can do more than just focus light to a point. As shown in *Figure 5-1*, they also can expand (or shrink) the bundle of parallel light rays that make up a laser beam. This is an important capability, because it can change the laser beam to the desired size. A telescope that serves this purpose is called a beam expander or collimator. Note that the same optics can expand or contract the laser beam, depending on which way the beam enters. Although we show a refracting telescope (one made with lenses) in our example, a reflecting mirror (which focuses with mirrors) also can serve as a beam collimator.

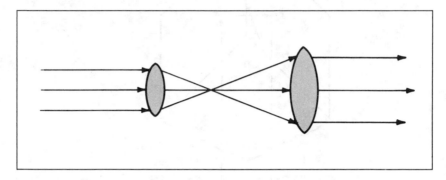

Figure 5-1. A telescope serves as a beam expander or collimator for a laser beam.

Retroreflectors

We saw in Chapter 2 that a flat mirror reflected an incident beam at an angle equal to the angle of reflection. Such a mirror would return a laser beam directly to its source if the beam struck it at precisely a right angle. That beam-returning capability is important for some laser measurements, but you can't be sure that the light will hit a flat mirror at a right angle.

It is possible to make a "retroreflector" that returns light precisely back in the direction from which it came by arranging three flat mirrors at right angles to one another—like the corner of a cube. This device is shown in *Figure 5-2*. Some retroreflectors are "hollow," made up of three front-surface mirrors. Others are prisms with the rear surfaces reflective.

Optical Aberrations

So far, we have assumed that simple optics do their job perfectly, bringing light of all wavelengths to a perfect focus. Life is not that kind. Optics suffer several kinds of aberrations which spread out the

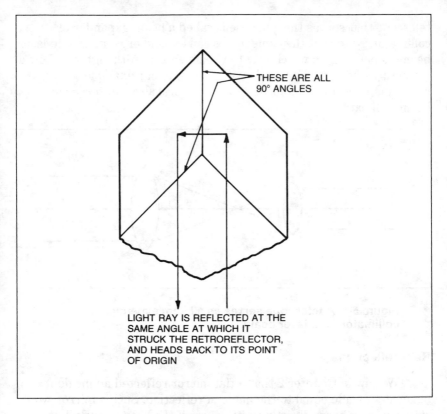

THESE ARE ALL
90° ANGLES

LIGHT RAY IS REFLECTED AT THE
SAME ANGLE AT WHICH IT
STRUCK THE RETROREFLECTOR,
AND HEADS BACK TO ITS POINT
OF ORIGIN

**Figure 5-2. A retroreflector returns light back in precisely
the direction from which it came.**

focal spot. The two most important for our purposes are chromatic
and spherical aberrations.

Chromatic aberration occurs in transmissive optics because
refractive index n is a function of wavelength. Recall that in Chapter
2, we learned that a positive lens has a focal length

$$f = 1 / \{(n-1) \times [(1 / R_1) + (1 / R_2)]\}$$

where,
R_1 and R_2 are the radii of curvature of the surfaces,
n is the refractive index of the glass.

This formula indicates that focal length changes with refractive index. The change is not large, but it can be significant. For example, in one common type of optical glass, n is 1.533 at 434 nm in the violet and 1.517 at 656 nm in the red. This means that the focal length in the red is 1.03 times that in the violet, or 3% longer. The effect is strongest near the edge of a lens or optical system, where white objects may seem to have a red fringe on one side and a blue fringe on the other.

Chromatic aberration is not a big problem with lasers that emit only one wavelength. However, compensation may be needed if the laser emits multiple wavelengths, or if the same optical system will be used by the human eye. The simplest way to compensate for chromatic aberration is to use compound lenses, made by gluing together components made of different glasses, with refractive indexes that change in different ways with wavelength. Careful selection of glass compositions can largely cancel the variations in refractive index with wavelength, although the correction is not perfect. Lenses with two or more elements designed to cancel chromatic aberration are called *achromatic* lenses.

Spherical aberration occurs because all parts of a spherical lens or mirror do not focus light to the same point. The outer parts have a slightly shorter focal length than the central zone. The difference is larger for more sharply curved surfaces. You can cope with the problem by using only the central part of the lens or mirror, or by picking optics with small curvature (and thus long focal length). An alternative is to make the surface with a nonspherical shape so it focuses all light rays at the same distance, no matter where they strike the surface. Such "aspheric" optics are harder to make than spherical surfaces, but work better, particularly at short focal lengths.

Coatings and Their Functions

Optical surfaces are normally coated with thin-film materials that enhance or suppress the reflection of light. The technology for making these coatings is complex, but the principles are fairly straightforward. We'll describe each basic type separately.

Antireflection Coatings

The theory of electromagnetic waves tells us that some light is always reflected when light passes from one material into another material with higher refractive index, such as from air into glass. The amount of reflection depends on the polarization of light and the refractive indexes of the materials. If the light leaving a material with refractive index n_1 strikes a transparent material with index n_2 at a normal angle, the reflectance is

$$\text{Reflectance} = \frac{[(n_2 / n_1) - 1]^2}{[(n_2 / n_1) + 1]^2}$$

If the light is entering glass, with $n_2 = 1.5$, from air ($n_1 = 1$), this gives a loss of about 4%.

At different angles, the reflectance differs for light polarized parallel to or perpendicular to the plane of incidence (the plane that contains the incident, transmitted, and reflected light rays). However, the overall dependence on refractive index remains. The larger the refractive index, the larger the losses caused by reflection.

A simple way to reduce this reflective loss is to reduce the difference in refractive index between air and the optical medium. You can't reduce the glass's index, but you can coat it with a lower-index material. The light must pass through two interfaces, but the reflective losses are much lower. If, for example, the coating has refractive index of 1.25, the reflectance at the air-coating interface is

$$\begin{aligned}\text{Reflectance} &= (1.25 - 1)^2 / (1.25 + 1)^2 \\ &= 1.23\%\end{aligned}$$

and the reflection at the coating-glass interface is:

$$\begin{aligned}\text{Reflectance} &= [(1.5/1.25) - 1]^2 / [(1.5/1.25) + 1]^2 \\ &= 0.83\%\end{aligned}$$

for a total reflectance loss of about 2%, half the level without the coating.

Interference Coatings

Much more complex effects are possible in multilayer "interference" or "dielectric" coatings. These coatings are made up of many alternating layers of two different materials, one with high refractive index and one with a lower index. The thicknesses of the layers are chosen so light reflected from one layer constructively (or destructively) interferes with that reflected from other layers. This lets the designer control transmission and reflection as a function of wavelength.

Proper selection of the layer materials and thicknesses lets coating designers meet demanding specifications. For example, they can make coatings that reflect or transmit only a narrow range of wavelengths. Such coatings can transmit (or block) a narrow range of wavelengths at a laser line. Thus, they can block white light from overhead lights while transmitting the red line of the helium-neon laser in the laser scanners used for automated checkout in supermarkets.

Interference coatings can give higher peak reflection or transmission (at the desired wavelength) than is otherwise possible. However, they have some practical limitations including their sensitivity to wavelength. Their transmission characteristics change with the angle at which light strikes them, because light incident at a different angle "sees" a different layer thickness, and thus experiences different interference effects. Tilt an interference coating that strongly reflects one color, and its apparent color will change with the viewing angle.

Optical Filters

Optical filters selectively attenuate some light that reaches them. The degree of attenuation depends on filter design. There are several kinds.

Spectral Filters

Wavelength-selective or spectral filters transmit some wavelengths and attenuate others. The degree of attenuation and selectivity vary widely. "Notch" filters transmit only a very narrow range of wavelengths; "bandpass" filters typically transmit a broader range. "High-pass" and "low-pass" filters transmit wavelengths shorter than (or longer than) a certain value.

Typically, interference coatings are used as filters for laser applications because of their good wavelength selectivity and the design flexibility they allow. Because they reflect rather than absorb rejected light, they can withstand reasonable laser powers.

Color Filters

Color filters serve a similar function to spectral filters in that they transmit certain wavelengths and attenuate others. However, they rely on the absorption of light by colored materials to select the transmitted wavelengths. This is fine for their major applications in photography. However, the wavelength selectivity is poorer than with interference coatings, a drawback for laser applications. In addition, because color filters absorb rejected wavelengths, they cannot block high powers without being damaged by heating.

Neutral-Density Filters

Neutral-density filters reduce or attenuate the light intensity uniformly at all wavelengths covered. That is, their absorption is essentially independent of wavelength, and they look gray to the eye.

The attenuation of a neutral-density filter is normally measured as optical density, defined as:

$$\text{Optical Density} = -\log_{10}(\text{Transmission})$$

Thus, if an optical filter transmitted 1% (0.01) of incident light, it would have an optical density of 2. Likewise, the fraction of transmitted light is given by reversing the formula:

$$\text{Transmission} = 10^{(-\text{Optical Density})}$$

Spatial Filters

A spatial filter blocks part of a beam and transmits the rest. A simple example and common type of spatial filter is a pinhole in a sheet of black-painted metal. It transmits the center of the laser beam, but not the outer portions. Often spatial filters are used to pick out the central, most uniform, part of a laser beam. Slits also are spatial filters.

Optical Materials

In Chapter 2, we saw that the way materials transmit light depends on wavelength. Materials we take for granted as transparent, such as air, water, and glass, are opaque at some infrared and ultraviolet wavelengths. Other materials that are opaque at visible wavelengths are transparent at other wavelengths. Reflectivity, likewise, varies with wavelength. As a consequence, different materials are used for optics in different parts of the spectrum.

What we call glass—an impure form of silicon dioxide, SiO_2—remains the most common transparent optical material for wavelengths in and near the visible. It is transparent, durable, and its technology is well-developed. We can make clear glasses with a range of refractive indexes sufficient to meet most optical needs. Quartz, a natural crystalline form of SiO_2 (silica), is used for some optics, as is purified silica. Other optical glasses contain dopants to raise the refractive index.

Clear plastics can be used for some visible optics. They are lighter than glass and have a lower refractive index. Their biggest practical advantage is the low cost of making molded plastic lenses. However, enough inhomogeneities remain to make plastic lenses unsuitable for many laser applications because of high light scattering, although they usually work well enough with white light.

A host of other transparent materials is used in other parts of the spectrum. *Figure 5-3* lists some of these materials and shows the wavelengths where they are transparent enough for optical applications.

CYLINDRICAL OPTICS

Not all lenses have spherical curvatures, but so far we have assumed that they all are circularly symmetric. That is, we assumed that a cross-section of the lens passing through its center would look the same no matter what direction you took the cross-section in. Light reaching such a lens at the same angle and the same distance from its center is always bent in the same way. Such lenses are fine for most purposes. You don't want a telescope to refract light differently if you rotate the lenses around the telescope axis.

However, there are cases where you don't want circularly symmetric optics. For example, the beams from semiconductor lasers are oval in shape, diverging more rapidly in one direction than in the direction perpendicular to it. To make the beam circular, you need an optical element that bends the light in one direction but not in the perpendicular direction. Such an element, shown in *Figure 5-4*, is called a cylindrical lens, because its surface looks like a piece of a cylinder.

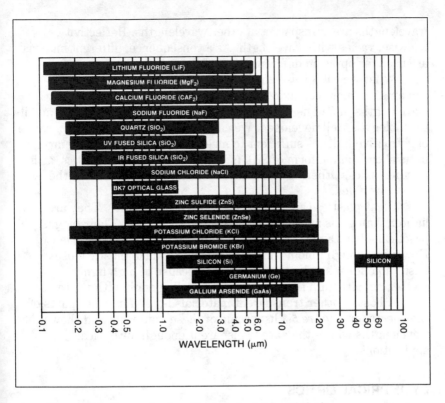

Figure 5-3. Important optical materials and their transmission ranges.

The action of a cylindrical lens depends on how you view it. If you look at a vertical plane in *Figure 5-4*, the lens cross-section is curved. Thus, the lens focuses light in that direction, bending light rays so they do not diverge as rapidly. However, if you look in a horizontal plane, the cylindrical lens is just a flat piece of glass, with no curvature and no focusing power. Thus the light rays go straight through, without being bent in that direction.

Cylindrical lenses normally are not used in imaging systems because—as you would imagine—they distort the image. However, they are important in some laser applications because they can collimate the asymmetrical beam from a semiconductor laser to make it look more like the better-behaved beam from a gas laser.

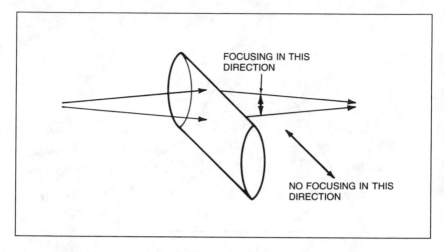

FOCUSING IN THIS DIRECTION

NO FOCUSING IN THIS DIRECTION

Figure 5-4. A cylindrical lens focuses light in one direction but not in the perpendicular direction.

DISPERSIVE OPTICS

Many times it is useful to be able to spread out or disperse different wavelengths of light. There are two types of optical components that can do the job—prisms and diffraction grating—which work on completely different principles.

Prisms

Prisms rely on the variation of refractive index with wavelength. When white light enters a prism, as shown in *Figure 5-5*, light toward the blue end of the spectrum is bent more than that at the red end, because glass has a higher refractive index at blue wavelengths. If light left the prism through a face parallel to the surface through which it entered, refraction at the exit face would cancel refraction at the entrance, and all wavelengths would emerge at the same angle. However, because the prism's output face is at an angle to the input face, different colors emerge at different angles, forming a spectrum.

Diffraction Gratings

A row of parallel grooves etched in a piece of glass or plastic also can spread out a spectrum. This effect is caused by the scattering

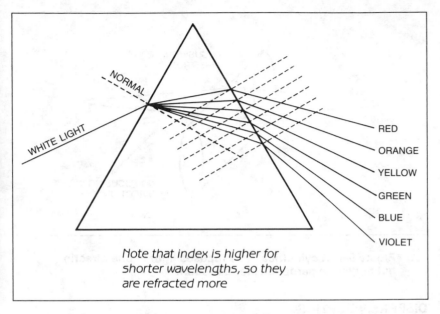

Note that index is higher for shorter wavelengths, so they are refracted more

Figure 5-5. A prism forms a spectrum by bending different wavelengths at different angles

of light from the different grooves. Light of all wavelengths is scattered at all angles. However, at many angles, light adds constructively only at one wavelength; all the other wavelengths add destructively, or interfere, reducing their intensity to zero. In the range of angles where the grating spreads out a spectrum, the result is a gradual change in wavelength with the angle, just as from a prism. The more grooves in the grating, the more the grating concentrates light in particular directions. Gratings can be transmissive or reflective, but reflective gratings are more common in optical instruments.

The way a reflective grating scatters light is described by the grating equation:

$$d(\sin i + \sin \theta) = m\lambda$$

where,

d is the grating spacing,
i is the angle of incidence on the grating (measured from the normal),

m is an integer indicating the order of the spectrum produced,
v is the angle of the scattered light to the normal,
λ is the wavelength.

If the input beam is at a normal angle to the grating, the equation reduces to

$$d(\sin \theta) = m\lambda$$

If we know the wavelength and the angle of incidence, we can convert this equation into a form suitable for calculating the emerging angle *θ*.

$$\theta = \arcsin[(m\lambda / d) - \sin i]$$

The presence of the factor *m* indicates that a diffraction grating produces more than one spectrum. For *m* = 0, we have undiffracted light reflected directly from (or transmitted directly through) the grating, without any dispersion (mathematically, multiplying by zero cancels the wavelength term). For *m* = 1, we have the first-order or strongest spectrum. At a larger angle, we have the second-order spectrum, formed by the *m* = 2 term. The intensity decreases as the value of *m* increases, so for most practical purposes we want the first-order spectrum. Note, however, that higher-order spectra have more dispersion because the wavelength is multiplied by *m*.

Spectroscopes and Other Instruments

Either diffraction gratings or prisms can be used to separate light of different wavelengths in optical instruments. One such instrument is the spectroscope, which projects different wavelengths of light at different angles, using a prism or grating. The spectroscope does not contain an internal light source.

A monochromator does contain an internal light source as well as a prism or grating to spread out the spectrum. In its simplest form, the prism or grating spreads out the spectrum onto a screen. A movable slit selects one color from the spectrum, and transmits that narrow range of wavelengths to the outside world. The viewer sees the monochromator emitting a narrow range of wavelengths, which is called "monochromatic" although strictly speaking it contains more than one wavelength.

5

FIBER OPTICS

Light waves normally travel in straight lines, but optical fibers can guide them along curved paths. The main use of optical fibers is to carry communication signals over long distances, but they also can be used in other types of optical systems.

A simplified view of the working of an optical fiber is shown in *Figure 5-6*. Light is transmitted by the core, which has a refractive index n_1 and is surrounded by a cladding layer with a lower refractive index n_2. If light in the core strikes the interface with the cladding at a glancing angle, it is reflected back into the core, because of the phenomenon of total internal reflection that we discussed briefly in Chapter 2. Light striking at larger angles leaks into the cladding layer and ultimately can escape from the fiber itself.

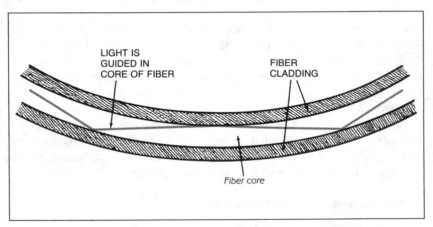

Figure 5-6. Light guiding in an optical fiber.

We should stress that this is a simplified view of how light travels through optical fibers. The subtleties are treated in more detail in a companion volume, *Understanding Fiber Optics* .* The most important point from the user's standpoint is that optical fibers guide light.

Fibers are made of glass or plastic that can transmit visible, near-infrared, and near-ultraviolet light. Plastic fibers transmit

Understanding Fiber Optics, (Cat. No. 27066), by Jeff Hecht, Howard W. Sams & Company, 1987.

visible light best, and are more flexible than glass. However, glass fibers have lower loss—particularly at near-infrared wavelengths of 1.0 to 1.6 micrometers—and thus are preferred for long-distance communications.

Single fibers usually are used for communications, where they carry signals generated by semiconductor lasers over long distances. Fibers are bundled together to "pipe" light into hard-to-reach places (including inside the human body), or to carry images from point to point. Fiber bundles may be made rigid (by melting the fibers together) or flexible (by leaving individual fibers loose in a housing). Fibers or fiber bundles are used to deliver laser light for some applications in medicine and industry.

The technology for making fibers from plastic or silica glass is well-developed. However, it is hard to make fibers from other optical materials. Much work is underway on fibers for wavelengths longer than 2 micrometers for potential applications with carbon-dioxide lasers and other types in medicine and industry. However, serious problems remain to be overcome with the materials.

POLARIZING OPTICS

Polarization indicates the direction of the oscillating electric field that is part of a light wave. As we mentioned earlier, you can divide light into two orthogonal polarizations, with electric fields perpendicular to each other. The simplest way to think of polarization is as a vector in the X-Y plane, as shown in *Figure 5-7*. The vector goes from the origin (0,0) to a point on the plane, and measures the amplitude and direction of the electric field. Each polarization vector is the sum of two perpendicular vectors, one in the X direction, the other in the Y direction. Those two vectors are the horizontal and vertical polarization components.

The X and Y directions are not absolute. If you rotate the two axes, you still can break the polarization vector into two perpendicular components aligned along the two new axes. The lengths of the two component vectors will differ, but their sum will remain the same.

Polarizers separate light into its polarization components. Simple dichroic film polarizers (like those used for sunglasses) absorb light polarized along one axis, and transmit light polarized along the other axis. For example, a polarizer might transmit light polarized

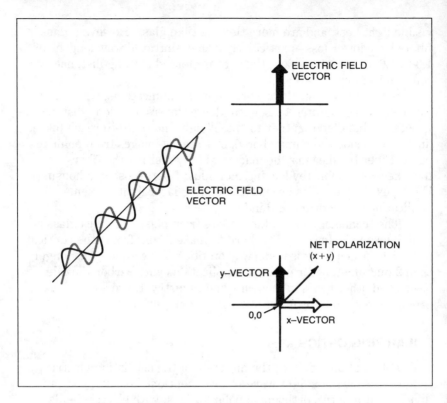

Figure 5-7. Direction of polarization as a vector.

along the Y axis in *Figure 5-7* and absorb light polarized along the X axis. Polarizers used with lasers normally work by other means, and separate light into two polarization components. Many rely on so-called "birefringent" materials, which have different refractive indexes for light of different polarizations. Others reflect light of certain polarizations.

Light is said to be linearly polarized if its electric field continually points in the same direction. If the polarization gradually changes with time but the amplitude is constant, the light is called "circularly polarized", because the polarization vector draws a circle in the X-Y plane. If the polarization vector changes both in direction and length, it draws an ellipse in the X-Y plane, and the light is called elliptically polarized.

Light is unpolarized if it is the sum of light of all different polarizations. (Random polarization—which might seem to be the same thing—is different, meaning light in which the direction of polarization changes continually in time.) Laser light may be polarized or unpolarized.

Polarization Retarders and Rotators

Because their refractive indexes depend on the polarization of light, birefringent materials can serve as polarization retarders or rotators. To understand how this works, suppose an unpolarized beam hits such a material, and consider what happens to the two polarization components. If the vertical polarization component sees a refractive index n, then the horizontal polarization component sees an index $n + \Delta$, where Δ is a small fraction of n (for convenience, we'll assume it's a positive number). Because the horizontally polarized light experiences a larger refractive index, it takes longer to travel through the material than the vertically polarized light. Depending on the application, this is called retarding or rotating the polarization.

In practice, polarization retarders are made with a thickness selected to retard one polarization behind the other by a quarter, a half, or a whole wavelength. Half- or quarter-wave plates can change the polarization of linearly polarized beams in various ways. For example, if the polarization of a linearly polarized input beam is at angle θ from the principal plane of the retarder, the polarization of the output beam is rotated by 2θ. If an input beam is linearly polarized at 45 degrees to the crystalline axis of a quarter-wave plate, the output beam is circularly polarized.

Polarization Analyzers

When polarizers are used in front of instruments, to separate light of one polarization from other polarizations for measurements, they are called "polarization analyzers."

BEAMSPLITTERS

One of the most important optical components used with lasers is the beamsplitter. It does what its name says, splitting a beam into two parts. Part of the light is transmitted, and part reflected. Normally, the beamsplitter is at an angle to the incident beam, so the reflected light goes off at an angle, while the transmitted beam passes right through, as shown in *Figure 5-8*. (You need not align the beamsplitter so the reflected beam is at a right angle to the incident and transmitted beam; this is just a convenient way to draw the beams.)

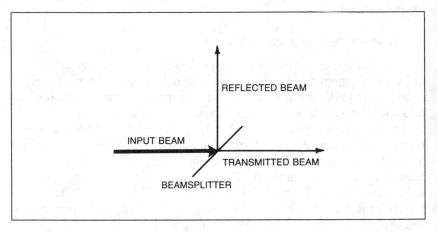

Figure 5-8. A beamsplitter.

Beamsplitters come in many types. Some reflect light of one linear polarization and transmit light of the orthogonal polarization. Others are insensitive to polarization. Some are multilayer interference coatings, and thus very sensitive to wavelength. Although some beamsplitters divide incident light equally into transmitted and reflected beams, others have different splitting ratios.

NONLINEAR OPTICS

So far we have talked mostly about linear optics, which do comparatively simple things to light. However, there also are nonlinear optics, which do more complex things impossible with linear optics. The theory of nonlinear optics is the stuff of three-page-long differential equations and PhD theses, but we can give you a basic idea here.

When we talk about linear optics, we mean devices in which the output is a simple function of the input:

$$\text{Output} = A \times (\text{Input})$$

Mathematically, the input and output are time-varying electromagnetic waves. If we ignore the phase of the light wave and consider only how the wave changes at a single point, the input wave can be represented by a formula like:

$$\text{Input} = Ke^{-i\omega t}$$

where,

K is a constant that indicates the amplitude of the electromagnetic wave,

i is the square root of -1 (indicating an imaginary number),

ω is the angular frequency (2π times the regular frequency ν),

t is the time.

Don't be afraid of the imaginary number; its main role here is to simplify the way we write the equation. We also can write the equation as:

$$\text{Input} = Ke^{-i\omega t}$$
$$= K[\cos(-i\omega t) + i\sin(-i\omega t)]$$

Ignoring the imaginary part of the equation lets us describe the light wave as a cosine wave. Although the cosine equation gives us a better sense of what the wave looks like, the exponential is easier to manipulate mathematically.

In nonlinear optics, the relationship between output and input are more complex because we have to consider other terms neglected before. Thus the equation looks like

$$\text{Output} = A(\text{Input}) + B(\text{Input})^2 + C(\text{Input})^3 + \text{ad infinitum}$$

where,

A, B, and C are constants or other expressions that include coefficients representing the strengths of various effects. In practice, the terms involving powers higher than the square of the input wave are usually unimportant.

Nonlinear optics often is broadly defined to include effects that do not depend purely upon the input wave. For example, external electric or magnetic fields can interact with a light wave's electromagnetic field to change the phase or polarization. You don't want to worry about the details (neither do I!), but you can get a useful overview of how they function.

Harmonic Generation

The simplest nonlinear optical effect to understand is harmonic generation, the production of light waves at integral multiples of the frequency of the original wave. The second harmonic is at twice the frequency of the fundamental wave, or half the wavelength. The third harmonic is at three times the frequency, or a third of the wavelength. In the laser world, the frequencies are very high, so the light waves normally are identified by wavelength, even though the process is still called harmonic generation. Thus laser specialists say they frequency-double the output of a 1064-nanometer neodymium laser to get 532-nanometer light, but they never mention the frequencies involved (2.83×10^{14} and 5.66×10^{14} hertz, respectively).

Harmonic generation makes sense if we look again at the nonlinear relationship of output to input,

$$\text{OUTPUT} = \text{A(INPUT)} + \text{B(INPUT)}^2 + \text{C(INPUT)}^3 + \text{ad infinitum}$$

If we substitute $Ke^{-i\omega t}$ for INPUT, we can see that in INPUT^2 term gives a factor of $e^{-i2\omega t}$, which has twice the frequency of the input wave. Thus we've doubled the frequency. Likewise, the INPUT^3 term gives a factor of $e^{-i3\omega t}$, which triples the frequency. (You can get similar results using trigonometric identities such as $\cos^2 \omega t = 0.5(\cos 2\omega t + 1)$—if you can remember the trigonometric identities.)

We can't forget the importance of the wave amplitude, which in our simple equation is given by K. If INPUT is $Ke^{-i\omega t}$, the INPUT^2 term is $K^2 e^{-i2\omega t}$—and depends not on K but on its square. Thus amplitude of the second-harmonic wave increases as the square of the field intensity at the fundamental frequency, not just with the field intensity. This means that the second-harmonic output rises much faster than input amplitude.

Despite its esoteric origins, harmonic generation is a practical tool in the laser world. There are few good laser materials around, so not all wavelengths are available readily. Harmonic generation makes more wavelengths available. Frequency-doubling of the neodymium laser to give 532 nanometer green light is the most important example, but there are many others.

In practice, most materials have very small nonlinear coefficients, and only a few materials are usable for second-harmonic generation. Special conditions must be met for the interaction between input and output beams. Because nonlinear effects are small, they usually appear only at high input power levels. As a result, it is easier to generate harmonics from short pulses with high peak power than from continuous-wave beams with a much lower steady power—even though the continuous-wave beam may deliver more energy over a much longer period.

Second-harmonic generation is most important in practice, but in some cases the third harmonic may be generated directly. Because the fourth-order nonlinear coefficient is much lower than the second-order coefficient, in practice the fourth harmonic is generated by frequency doubling the second harmonic.

Sum and Difference Frequency Generation

If you take another look at our nonlinear equation, you can see another possible nonlinear effect—generation of light waves at sum and difference frequencies. Suppose the input light is the sum of two waves at different frequencies, ω and ϕ.

$$\text{Input} = K_1 e^{-i\omega t} + K_2 e^{-i\phi t}$$

Then we see that the INPUT^2 term includes a factor $e^{-i(\omega+\phi)t}$, meaning that light is generated at the sum of the two frequencies. There also is a difference-frequency term, with frequency $\omega - \phi$.

Like harmonic generation, sum- and difference-frequency generation effects are weak in most materials, and require special conditions. They depend on the product of the two input intensities, $K_1 \times K_2$ in our example. That means that the output increases sharply as the combined power level increases.

Parametric Oscillators

In sum-frequency generation, light at frequencies ω and ϕ add together to produce the frequency $\omega + \phi$. Under certain circumstances, you can turn the equation around, and use a strong beam at $\omega + \phi$ to generate weak beams at ω and ϕ.

In a parametric oscillator, pump light at the frequency $\omega + \phi$ enters a nonlinear medium that is placed between two mirrors. The mirrors are reflective at one or both of the frequencies ω and ϕ, forming a resonant cavity comparable to a laser cavity. The nature and orientation of the nonlinear material determines what frequencies ω and ϕ are produced.

In practice, parametric oscillators are designed to produce one wave, at a frequency we'll call ω, as the signal; the second wave, which we'll say is at frequency ϕ is called the idler, and is discarded. Natural noise processes generate weak light at both ω and ϕ. Mixing this weak light at ϕ with the pump signal at $\omega + \phi$ generates a weak difference-frequency signal at ω. The resonance of one or both of the ω and ϕ waves in the cavity enhances the interaction in the nonlinear crystal, thus building up the signal wave, which can be tuned in wavelength by adjusting the parametric oscillator.

Parametric oscillators are not widely used, but they can be valuable for special purposes, particularly generating wavelengths that otherwise are hard to produce.

Raman Shifting

Another interaction usually considered nonlinear is a process called Raman shifting, after its discoverer, Indian physicist C. V. Raman. Most light reflected (or scattered) from atoms or molecules is unchanged in frequency. However, a small fraction of the incident light is changed in frequency, a process called Raman shifting.

The Raman shift is an absorption or emission of excess energy when the light wave is scattered from the medium. The shift occurs when the medium undergoes a vibrational transition, at one of the material's characteristic frequencies. If the material drops in vibrational energy, the Raman shift releases energy, adding energy to the scattered light wave and decreasing its wavelength. If the material increases its vibrational energy, it absorbs some energy from the light wave, which then is scattered at a longer wavelength.

Incident light at any wavelength can experience a Raman shift, although it shows up best at higher power levels. The amount of Raman shift (measured in energy units) is an inherent characteristic of the material, and is the same for the material at different wavelengths. Normally there are many possible Raman-shifted wavelengths.

Raman shifting is attractive for shifting laser wavelengths. One use is to shift the ultraviolet wavelengths produced by excimer lasers to longer or shorter wavelengths that may be more useful.

INTENSITY MODULATION

Many applications require changing or modulating power in a laser beam, either to make it vary with time or simply to switch it on or off. This can be done in several ways: by blocking the beam mechanically, by changing the electrical or optical power that drives the laser, or by passing the beam through a device with variable transmission characteristics. Let's look briefly at each approach.

Mechanical Modulation

Shutters and beam choppers mechanically block a beam. A shutter, like those on a camera, opens to let the beam through and closes to block light. Shutters are used as safety devices on many lasers, to block the beam when it is not needed, or when the laser may not be operated correctly. Control can be manual or automatic.

Shutters operate in a fraction of a second, but that is slow by laser standards. Mechanical beam choppers can modulate a laser beam much faster. A typical example is a rotating disk with holes or slots to let the beam through periodically. You can design shutters and beam choppers to block only part of a beam, but they do not precisely control the fraction of light transmitted (because of uncertainties in the beam position and the distribution of power in the beam), so they typically are used only to turn the beam off and on.

Direct Modulation

We control operation of light bulbs by turning the electricity off and on, and lasers can be controlled in the same way. Direct modulation often is the simplest and cheapest way to control laser power. However, this approach works better for some lasers than for others because of differences in internal operating conditions and response time.

Direct electrical modulation works best for semiconductor diode lasers. The devices are small, and respond very quickly to changes in drive current; specially designed lasers have been directly modulated at billions of pulses per second. Only a simple circuit is needed to control the drive current, although circuit complexity does increase with operating speed. An added advantage is that above laser threshold the light output increases in proportion to an increase in drive current, as shown in *Figure 5-9*.

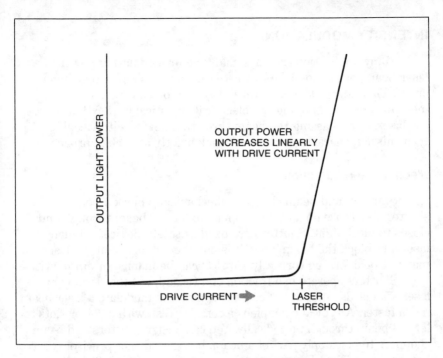

Figure 5-9. Output-current curve for a diode laser.

Gas lasers are not as well-suited for direct electrical modulation. They can be turned off and on, but they take much longer to stabilize—several minutes for some types. Because they need potentials of a thousand volts or more across the tube, modulation also is a more complex task electronically. In addition, their output power does not depend linearly on drive current or voltage. You might compare the difference between semiconductor and gas lasers to the difference between incandescent bulbs and fluorescent tubes. Incandescent bulbs turn off and on faster, and their light output can be adjusted with a dimmer; fluorescent tubes are harder to ignite and their brightness cannot be adjusted with a dimmer.

Direct Optical Modulation

We saw earlier that some lasers are pumped optically. Such lasers respond very quickly to changes in the pump light, so you can modulate them by changing the optical input. If a laser is pumped by a flashlamp or pulsed laser, its output rises and falls with pump-light intensity. This does not make direct modulation simple, however; it just shifts the problem of controlling output intensity to a different place—the pump source. Pulsing a flashlamp does produce a laser pulse from a flashlamp-pumped laser, but electronics must pulse the flashlamp (which again is slow by laser standards). Likewise, modulating the pump laser can control the output of an optically pumped laser, but you still have to adjust the pump laser output.

External Electro-Optical Modulation

Often the only practical way to modulate the intensity of a laser beam is with an external device that can be controlled to transmit varying fractions of the beam. The problem is finding a suitable device. While transmission differs among materials, the optical characteristics of a given material rarely change very much.

One type of external modulator relies on small changes in the refractive index caused by an electric field. We saw earlier that the refractive index of some materials depends on the polarization of light and the crystal orientation. Applying an electric field along the proper crystalline axis can change this birefringence.

When we talked about polarization rotators earlier, we learned that the rotation angle depended on the material's thickness and birefringence. For electro-optic crystals, the polarization angle depends not only on the material thickness but also on the electric field. Once the electro-optic material is chosen and cut to size, the only significant variable is the electric field, which thus controls the angle of polarization rotation. (We should note other factors, including temperature and pressure, can change birefringence slightly, but not as much as the electric field.)

Merely rotating the polarization of light does not change its intensity. *Figure 5-10* shows the construction of an electro-optic modulator. First a polarizing filter gives the beam a linear polarization. Then the crystal rotates the polarization. (The degree of rotation depends on the voltage applied across the electrodes.) Then the emerging light must pass through another polarizing filter, with its plane of polarization at the proper angle. Adjust the voltage across the crystal so it rotates the plane of polarization to be parallel to that of the output filter, and the transmitted beam is at its maximum intensity. Rotate it so it is perpendicular to the plane of the filter, and the beam will be blocked. In *Figure 5-10*, we see the output intensity somewhat reduced.

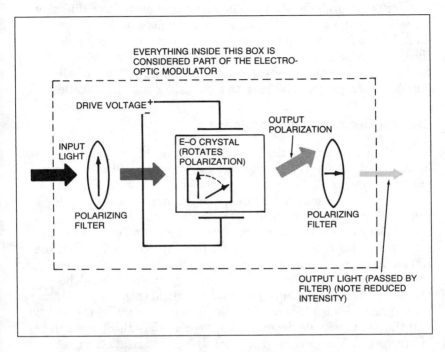

Figure 5-10. An electro-optic modulator, with an electro-optic cell mounted between crossed polarizers.

There are two types of electro-optic modulators: Kerr and Pockels cells. They rely on subtly different effects. The more common Pockels cell relies on the Pockels effect, in which the electric field is parallel or perpendicular to the crystal's optical axis. The change in birefringence is proportional to a constant (p) times the applied electric field (E), or pE. The constant is small; typical values of p are 10^{-10} to 10^{-11} meter per volt even for the best materials. The Kerr effect arises from alignment of molecules in a liquid or gas parallel to the electric field direction. It changes birefringence by an amount equal to the product of a constant K times the wavelength λ times the square of the electric field E or, $K\lambda E^2$. Kerr cells are fast, but they require higher voltages than Pockels cells to rotate the polarization by the same amount.

Both types of electro-optical modulators are expensive, costing hundreds of dollars and up. They also have other limitations, including the need for bias of tens—or more typically, hundreds—of volts.

External Acousto-Optic Modulators

A laser beam also can be modulated by interactions with sound waves in a solid. As an acoustic wave passes through a solid, it alternately raises and lowers the pressure, which, in turn, changes the refractive index. Light waves passing through the material see these variations in refractive index as a series of layers, like the multilayer dielectric coating we described earlier. Alternatively, the refractive-index variations can be seen as lines in a diffraction grating.

Interference and diffraction effects combine to scatter light from the "acoustic grating" at an angle that depends on the ratio of the wavelengths of light and sound in the medium. The stronger the acoustic waves, the more light is scattered or deflected from the main beam.

This effect can modulate light if we look only at the main beam and throw away the scattered light. (Later we will see other uses for the deflected beam.) Suppose we add a spatial filter to transmit the main beam but block all other light. Then the light intensity will drop as intensity of sound waves increases—modulating the beam intensity.

Like electro-optic modulators, acousto-optic modulators cost hundreds of dollars and up. The two types are somewhat competitive.

BEAM SCANNERS

Many applications require scanning a laser beam back and forth, much as your eyes scan this page or as an electron beam scans the screen of a television tube. There are several major approaches.

Perhaps the simplest concept is the rotating mirrors. This is a reflective polygon that spins rapidly around a central axis. Each mirror face turns as the laser beam strikes it, scanning the reflected beam across an angle dictated by the rule that the angle of incidence equals the angle of reflection. Each new facet repeats the scan. Holographic scanners are similar in concept, but rely on rotating holographic optical elements to direct the light.

The mechanical motion of mirrors also is the basis of resonant and galvanometer scanners. However, these scanners twist flat mirrors back and forth over a limited angular range, so the laser beam always strikes the same flat mirror surface. As the mirror swings back and forth, it sweeps the laser beam across a line.

As we mentioned in describing acousto-optic modulators, the interaction between a light wave and a sound wave can deflect some light passing through certain materials. In modulators, we use the undeflected beam, but in beam scanners we use the deflected light. The angle at which the deflected light emerges is proportional to the ratio of optical to acoustic wavelengths; the strength of the deflected light depends on the acoustic power. Thus, you can make the light scan a pattern by controlling the acoustic frequency driving the scanner. Unlike mechanical scanners, acousto-optic deflectors are solid-state devices with no moving parts.

It also is possible to make electro-optical deflectors, which like acousto-optic scanners are solid-state devices, but depend on electro-optic interactions. However, these are rarely used.

CONTROLLING LASER PULSE CHARACTERISTICS

We saw earlier the importance of the variation of laser output with time. It is often useful to concentrate laser energy into short intervals. Like focusing a broad beam down to a tiny spot, this concentrates the energy in a single place and time. There are three primary tools: Q switches, cavity dumpers, and modelockers.

Q Switches

Every resonant laser cavity has a characteristic quality factor or Q that measures internal loss. The higher the Q, the lower the loss.

Loss is not necessarily bad, because a high-loss cavity can store more energy than one with lower loss. Suppose, for example, you block one laser mirror, then excite the laser medium. The excitation can produce a population inversion, but the laser cannot oscillate because the mirror does not reflect light. If you suddenly unblock the mirror, the population inversion will be much larger than needed for laser action. Stimulated emission will quickly drain the stored laser energy from the cavity in a short pulse with peak power much higher than the laser could otherwise produce. You can think of a Q switch as a device that quickly switches from absorptive to transmissive, suddenly reducing cavity loss.

The length of a Q-switched pulse depends on output-mirror reflectivity R and on the time t it takes laser light to make a round-trip through the laser cavity:

$$\text{Pulse Length} = t \,/\, (1-R)$$

Typical pulse lengths are around 10 nanoseconds. The round-trip time equals twice the cavity length L divided by the speed of light in the laser medium c_{med}, which itself equals the speed of light in vacuum c divided by the laser medium's refractive index n. Put this together and you have a more useful equation for Q-switched pulse length

$$\text{Pulse Length} = 2Ln \,/\, c(1-R)$$

A Q switch can change the cavity Q once to produce a single laser shot, or repeatedly to generate a series of pulses at a specified interval, depending on how it is driven. Both electro-optic and acousto-optic modulators can serve as Q switches. Other Q switches incorporate rotating mirrors. When the mirror face is not perpendicular to the laser axis, it does not reflect light back to the output mirror and the Q is low; when the mirror face is perpendicular to the laser axis, the Q is momentarily high, and the laser generates an intense pulse of light. *Figure 5-11* shows operation of a Q switch.

Not all Q switches require active control. A "passive" Q switch can be built using a dye that normally absorbs light. After the dye absorbs a photon, it stays a while in an upper energy level, where it can't absorb another photon. Strong enough stimulated emission from the laser medium can quickly raise all of the dye atoms to the excited state, where they can't absorb more laser photons. Once that happens, we say the dye has been "bleached"—that is, it has become transparent because no molecules are left to absorb light. The abrupt change of the dye from absorbing to transmitting light raises the cavity Q by exposing a mirror behind it—and thus generates a Q-switched pulse.

Q switching is widely used, but not all lasers can be used with Q switches. Q switching works only if the laser can accumulate energy in excited-state atoms or molecules for a period longer than the Q-switched pulse. Another way of seeing this is to say that the excited

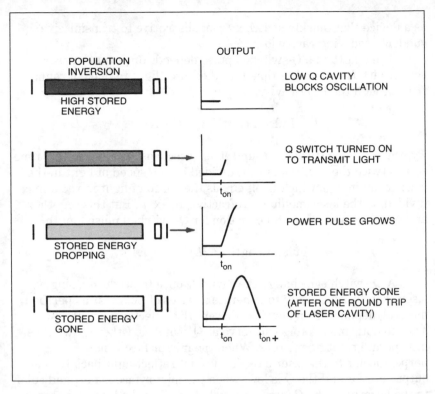

Figure 5-11. Q-switching generates a laser pulse.

state must have a long spontaneous-emission lifetime. Although many lasers do store energy in that way, there are some exceptions.

Cavity Dumping

Like Q switching, cavity dumping works by releasing energy stored in the laser cavity while it is kept from oscillating. However, the details are quite different. In essence, a cavity-dumped laser normally has two fully reflective mirrors, one on each end of the cavity. The laser normally operates with this high-Q cavity, building up a steady power inside the cavity. However, the power cannot escape as long as the two fully reflecting mirrors are at the ends of the cavity.

How do you get the power out? You can put another mirror in the laser cavity that aims the beam out of the cavity, instead of back at the other resonator mirror. This would dump all the energy circulating in the laser cavity in a single pulse, lasting only as long as the light took to make a round-trip of the laser cavity. This means that the length of a cavity-dumped pulse is twice the cavity length divided by the speed of light in the medium. For a 30-centimeter gas laser, this would be two nanoseconds, significantly shorter than a Q-switched pulse.

Cavity dumpers don't actually use pop-up mirrors. Instead, they rely on beam deflectors or other components inside the laser cavity that briefly deflect the beam outside of the laser. Cavity dumping is not used as often as Q switching, and the pulses it generates have lower energy than Q-switched lasers. However, lasers that cannot be Q-switched can be cavity dumped.

Modelocking Pulses

The shortest laser pulses—lasting on the order of a picosecond (10^{-12} second)—are generated by a process called "modelocking." The idea, from a theoretical physicist's point of view, is to lock together many longitudinal modes so a laser simultaneously oscillates in phase on all of them. You get this control by inserting a special optical element in the laser cavity that makes all the modes oscillate together.

It may be simpler to visualize the process by imagining that it is a clump of photons, not modes, that are locked together. The clump bounces back and forth between the laser mirrors. The modelocking element in the laser cavity "opens up" to transmit light each time the photon clump passes through. Each time the photon clump hits the output mirror, a short pulse escapes. Outside the laser, we see many short pulses, one emitted each time the clump of photons makes a round-trip of the cavity. Modelocked pulses are very short, but during those very short pulses the peak power can reach a very high level.

The length of a modelocked pulse depends on the range of wavelengths (or, equivalently, frequencies) emitted by the laser. The wider the frequency range $\Delta\nu$, the shorter the pulse can be. The lower limit on pulse duration is roughly:

Minimum Pulse Duration = 1 / Frequency Range

or

$$t = 1 / \Delta\nu$$

Pulses of this minimum length are sometimes called "transform-limited," because the relationship is considered a transform between time and frequency domains.

Note that the natural repetition rate of a modelocked laser is the time it takes light to make a round-trip of the laser cavity. That isn't long—light takes only about 2 nanoseconds to make a round-trip of a 30-centimeter (approximately one-foot) long laser. Thus a 30-cm modelocked laser would generate 500 million pulses a second (a 500-megahertz repetition rate) much higher than the repetition rates of cavity dumping or Q switching.

Combining Pulse Controls

It is possible to combine two or more of these pulse control techniques. For example, combining cavity dumping and modelocking can help reduce the repetition rate of modelocked pulses, and ease measurement requirements or help isolate effects of successive pulses.

POWER AND ENERGY MEASUREMENT

Special instruments, devices, and terms are used in measuring laser power and energy. While we don't have room to give a comprehensive overview, we can introduce basic ideas of light detection and measurement.

Light Detection

The most common way to measure light is to first convert it to electricity, then measure electrical current, resistance, or voltage. Various light detectors can convert optical signals into electronic form. The most common are semiconductor devices, in which incident light raises an electron from the valence band to the conduction band, and also creates a "hole" in the conduction band which functions as a current carrier. An older type is the "photoemissive" detector, a vacuum tube in which light frees electrons from a metal surface, and those electrons are collected by a positively biased electrode. Photomultipliers are photoemissive tubes with several stages that amplify the electrical signal. (Originally, "photodiode" meant a vacuum photoemissive tube with two electrodes—one that emits electrons and the other that collects them—but the word now is used for semiconductor detectors.) There also are some detectors that monitor light in other ways.

Each type of detector is sensitive to a limited range of wavelengths, depending on its composition. A small sampling is listed in *Table 5-1*. Many types are available, with the choice depending on wavelength range, sensitivity to light, the speed with which they can respond to signals, durability, operating requirements, and cost.

Table 5-1. Wavelength Ranges of Detectors

Type	Material	Wavelengths (nm)
Photoemissive	Potassium-cesium-antimony	200–600
Semiconductor	Silicon	400–1000
Semiconductor	Germanium	600–1800
Semiconductor	Indium arsenide	1500–3000
Semiconductor	Lead sulfide	1500–3300
Semiconductor	Lead selenide	1500–6000

The electrical output from detectors can be used for measurement or other purposes. For example, electrical signals from detectors in optical communication systems are processed by other electronic circuitry so they can drive electronic equipment such as computers or telephone switching systems.

Sometimes the term "detector" can be used in a broader sense to include light sensors that do not produce electrical output. Simple examples include photographic film and photochromic materials, in which light changes how the material looks to the eye.

Radiometry, Photometry, and Light Measurement

The measurement of light is a specialized art often divided into two fields, radiometry and photometry. Radiometry measures the power and energy contained in electromagnetic radiation, regardless of wavelength. Photometry measures only light visible to the human eye, with the contribution of each wavelength weighted according to the eye's sensitivity. Thus, photometry ignores infrared wavelengths invisible to the human eye, and counts photons of green light (to which the eye is very sensitive) much more strongly than red or violet light (to which the eye does not respond as strongly). Unfortunately, some people are sloppy with that terminology, but you should remember the difference.

The eye almost never looks directly at laser light (see Appendix A on safety), so virtually all laser measurements are made in radiometric units. *Table 5-2* summarizes these units and lists common symbols for them, with brief descriptions of their meanings.

Table 5-2. Radiometric Units

Quantity and Symbol	Meaning	Units
Energy (Q)	Amount of light energy	joules (J)
Power (P or Φ)	Flow of light energy past a point at a particular time (dQ/dt)	watts (W)
Intensity (I)	Power per unit solid angle	watts/steradian
Irradiance (E)	Power incident per unit area	watts/cm^2
Radiance (L)	Power per unit angle per unit projected area	W/steradian-m^2

Note that power (sometimes called radiant flux) is a measure of the rate at which electromagnetic energy flows by a point. It is inherently a function of time, the derivitive of the energy:

$$\Phi = dQ\,/\,dt$$

or

$$\text{Power} = d(\text{Energy})\,/\,d(\text{Time})$$

The unit for measuring power, the watt, equals one joule per second.

Some commercial instruments also are calibrated to measure power in decibels. The decibel (dB) is a relative unit that gives the ratio of two power levels, P_1 and P_2:

$$\text{Power Ratio (dB)} = 10 \log_{10}(P_1 / P_2)$$

Sometimes powers may be measured in decibels relative to some predefined level, typically one milliwatt (1 mW) or one microwatt (1 μW). A positive number indicates that the measured power level is above the comparison, a negative number indicates it is lower.

Pulse Duration & Spectral Measurements

Pulse duration and light wavelength are two other important quantities that sometimes must be measured while working with lasers. Both require special measurement instruments.

The simplest way to measure pulse length is to feed the output of an electronic detector into an oscilloscope. However, this runs into instrumental limitations if the pulse lengths decrease to the nanosecond realm. Sophisticated sampling oscilloscopes can measure nanosecond pulses accurately, but more complex techniques are needed to measure picosecond and subpicosecond pulses. Such techniques are outside the scope of this book.

The simplest way to analyze the spectrum of light is to spread it out with a prism or diffraction grating. This method is adequate for some simple measurements, but when more quantitative measurements are needed this method may have to be combined with electronic techniques. One valuable approach is to spread out the spectrum on a linear array of light sensors, then measure the amount of light reaching each sensor. With proper calibration, the measurements from individual sensors indicate how much light is present at particular wavelengths.

MOUNTING AND POSITIONING EQUIPMENT

Walk into any well-equipped laser laboratory, and the first thing you are likely to notice is a massive table, probably painted deep black, such as the one shown in *Figure 5-12*. This behemoth is called an optical table. On it you will find an array of special mounting equipment that holds lasers, lenses, and other optical components.

Figure 5-12. Optical table, with mounts. *(Courtesy Newport Corp.)*

Optical tables and mounts are not optics per se, but they are used with optics and lasers. Optical tables and massive linear rails called optical benches serve as foundations. Many are mounted on special shock-absorbing legs that isolate them from vibrations in the room. This vibration isolation is essential for holography and sensitive measurement because small effects, such as a person walking through the room or a truck passing on the street outside, can cause vibrations comparable in size to the wavelength of light. The tables are expensive, but alternatives exist if you're working on a low budget in a home or school laboratory. Most books on making your own holograms describe how to isolate optical setups from vibrations by mounting them in a sandbox.

An optical table or bench also serves as a firm foundation. Typically, threaded holes are drilled in the surface of optical tables to accommodate screw-in mounts. The edges of optical rails and benches mate with optical mounts that hold components in place. The mounts themselves are made to hold standard lenses and optics firmly in place.

In addition to fixed mounts, tables, and optical benches, optical laboratories also use special positioning equipment to hold and move optical components. The motion may be driven manually or mechanically, and some equipment incorporates computer controls. Precision is crucial because laser and optical systems can be very sensitive to small misalignments.

The technology of optical mounts and positioning equipment is mundane compared to lasers. You can live without the best-quality equipment (indeed, if your budget is tight, you may not have any choice). However, good mounting and positioning equipment can make life much easier.

EMERGING TECHNOLOGIES

Some potentially important optical technologies remain in the laboratory. Three of them deserve brief mention because you will be hearing about them in the years to come: Integrated optics, adaptive optics, and phase conjugation

Integrated Optics

The basic idea behind integrated optics is something like that behind integrated electronic circuits. Today's lasers and optics are discrete, each made from a separate chunk of material. Electronics used to be that way, too, with circuits made of separate transistors, diodes, resistors, and capacitors. Suppose we could integrate the function of an entire optical system—laser, lenses, mirrors, and modulator, for example—into a single chunk of material. Then we would have integrated optics. Such integrated optical circuits are attractive for applications such as communications and signal processing that require manipulating low-power beams of light.

Researchers have been working on integrated optics for many years, and have achieved modest success. However, they also face some stubborn problems. Important difficulties come from the nature of the materials themselves. For example, you can't make lasers from the materials most useful for making modulators. The interactions that affect how light travels through a "waveguide" in an integrated optical circuit are so weak that the light must travel through a comparatively long channel to be switched between output ports—meaning that integrated optics must be much larger than their electronic counterparts. Although the technology remains promising, it is far from ready for routine use.

Adaptive Optics

One of our implicit assumptions about optics is that they are fixed and unchanging. We think of a mirror or lens retaining the same curvature throughout its usable lifetime—and we might consider its useful lifetime ended if something did change its shape. Suppose, however, that optical components could change their shape to meet changing conditions.

Interest in making adaptive optics comes from the effects of the atmosphere on light. Air currents and turbulence bend light rays. We see this at night as the twinkling of stars. Astronomers with large telescopes find that it spreads starlight, limiting resolution of the largest telescopes to about one arcsecond. Such fluctuations also affect efforts to send beams from high-power laser weapons through the atmosphere.

The idea of adaptive optics is to bend the surface of a mirror to compensate for atmospheric effects. The changes must be made continually, because the atmosphere changes continually. The optics first sense what is happening in the atmosphere, then the surface of the mirror is bent to compensate for those atmospheric effects. If this works properly, it can let astronomers focus the image of a star to a point, or military engineers send a high-energy laser beam to a small spot on the target. (Although the Pentagon has paid most of the bills so far, the military problems are the toughest because high-power laser beams themselves generate some side effects that are hard to correct.)

Phase Conjugation

Another idea being studied to correct for atmospheric effects on light propagation is called phase conjugation. In a sense, this is a step beyond the retroreflector that we described earlier.

The retroreflector sends light rays back in the direction that they came. Phase conjugation does somewhat the same thing in a different way. It generates what is called a "conjugate" of an incoming light wave. The conjugate wave is essentially a reversed version of the original wave—the wave is at the same phase and shape, but travelling in a different direction, as shown in *Figure 5-13*. This phase-conjugate wave travels in the same path as the incoming wave, but in exactly the opposite direction. If the incoming light all came from a small laser spot, the phase-conjugate wave will return to that same spot.

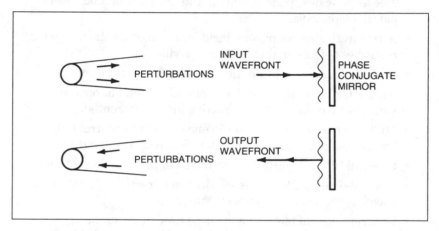

Figure 5-13. Generation of a phase-conjugate wave.

That may sound exactly like retroreflection. However, it is actually a time-reversed version of the original wave, reconstructing the light's phase and amplitude. Because of the way the phase-conjugate wave recreates the original wave, it automatically compensates for things that happened to the laser beam. In *Figure 5-13*, the incoming light waves experience perturbations that cause them to spread out. The phase conjugate wave experiences the same perturbations, but in the opposite direction, as if it's going backwards in time. Thus the spread-out light in the phase-conjugate beam is pushed back together and focused on the laser source.

Phase conjugation remains in the laboratory, where researchers are trying to find better materials to generate the conjugate beam. Meanwhile, laser specialists are investigating possible applications.

5

What Have We Learned?

- Passive optics always do the same things; active optics can change what they do to light.
- A positive lens or concave mirror focuses parallel light rays to a point.
- Telescopes can expand or shrink the diameter of a bundle of parallel light rays.
- A retroreflector returns an incident beam to its starting point, regardless of the angle at which it strikes.
- Optics suffer from chromatic and spherical aberration.
- Antireflection coatings reduce reflective losses at optical surfaces by reducing the refractive-index differential.
- Multilayer coatings use interference effects to control light transmission and reflection as a function of wavelength.
- Spectral filters transmit some wavelengths and block others.
- Neutral-density filters absorb the same fraction of light at all wavelengths in their operating range.
- Different optical materials must be used at different wavelengths. Glass is useful only in and near the visible range.
- Cylindrical optics refract light in one direction but not in the perpendicular direction.
- Prisms spread light out by wavelength because of the way refractive index varies with wavelength.
- Diffraction gratings are rows of parallel grooves which scatter light by interference effects to form multiple spectra.
- Fiber optics guide light in a core that has higher refractive index then the surrounding cladding.
- Polarization is the direction of the electric field in a light wave. We can think of it as a vector. It can be retarded or rotated by birefringent materials.
- Beamsplitters separate an input beam into two parts.
- The output of nonlinear optical components does not depend linearly on the input, permitting special optical effects.

5

- Nonlinear optical materials can produce light waves at multiples of the input frequency.
- Amplitude of a second-harmonic wave increases as the square of the fundamental intensity.
- Only a few materials have high nonlinear coefficients.
- Nonlinear interactions can generate light waves at sum and difference frequencies.
- A parametric oscillator can generate tunable light at longer wavelengths than the input, by effects related to sum-frequency generation.
- Raman shifting changes the wavelength of light scattered from certain materials.
- Beams can be blocked by mechanical shutters and beam choppers.
- Direct modulation turns laser excitation off and on to control output power. It works best for semiconductor lasers.
- External electro-optic modulators change polarization to modulate light intensity.
- Interactions with acoustic waves in a solid can modulate a laser beam.
- Mechanical devices, or electro-optic or acousto-optic devices can scan laser beams.
- A Q switch controls cavity losses to generate short, intense pulses. Q switching works only for laser media that can store energy.
- A cavity dumper "dumps" energy from a high-Q laser cavity.
- Modelocking generates ultrashort pulses by effectively making a clump of photons circulate back and forth through the laser cavity.
- Length of a modelocked pulse depends on laser bandwidth.
- Light usually is converted to electricity so it can be measured.
- Radiometry is the measurement of electromagnetic power at all wavelengths.
- Photometry is the measurement of only the light visible to the human eye.

5

- Power is a measure of the rate of flow of electromagnetic energy.
- Pulse duration and wavelength are important quantities when working with lasers.
- Optical tables support optical equipment and isolate it from vibrations
- Integrated optics combine two or more optical functions in a single device.
- Technology for integrating optical components together is still developing.
- Adaptive optics are in development that change their surface shape to correct for changing conditions in the atmosphere.
- Phase conjugation generates a time-reversed version of an input light wave.

WHAT'S NEXT?

In Chapter 6 we will start looking at specific types of lasers. Our first family of lasers will be gas lasers.

Quiz for Chapter 5

1. A double-convex lens with two equal radii of curvature of 20 centimeters has refractive index of 1.60 at 400 nm and 1.50 at 700 nm. What is the difference between the lens' focal lengths at those two wavelengths?
   a. 400 nm focal length is 3.33 cm shorter
   b. 400 nm focal length is 0.33 cm shorter
   c. 700 nm focal length is 0.033 cm shorter
   d. 700 nm focal length is 1 cm shorter
   e. No difference

2. Silicon has a refractive index of 3.42 at 6 micrometers in the infrared. What is the reflective loss for 6-μm light incident from air normal to the surface?
   a. 5%
   b. 10%
   c. 20%
   d. 30%
   e. 40%

3. For the silicon sample in Problem 2, what is the reflective loss if a coating with refractive index of 2 is applied on the surface? (Remember that you can't just add the reflective losses. You must multiply the fractions of transmitted light to get the total transmitted light to assess overall reflective loss.)
   a. 5%
   b. 10%
   c. 15.4%
   d. 17.2%
   e. 20.5%

4. What type of filter would you use to block light from a laser, but let light from other sources through?
   a. Neutral-density filter
   b. Interference filter
   c. Color filter
   d. Spatial filter
   e. Any of the above

5

5. At what angle (from the normal) will a diffraction grating with one-micrometer spacing scatter 500-nm light if the input light is incident at a normal angle? Calculate for the first-order spectrum (m = 1).
   a. 0.5 degree
   b. 5 degrees
   c. 30 degrees
   d. 60 degrees
   e. None of the above

6. What is the wavelength of the fourth harmonic of the ruby laser (694 nm)?
   a. 2776 nm
   b. 1060 nm
   c. 698 nm
   d. 347 nm
   e. 173.5 nm

7. Raman shifting does what to input light?
   a. Modulates its intensity.
   b. Doubles its frequency
   c. Changes its wavelength by a modest amount
   d. A and B
   e. None

8. Which type of laser is the simplest to modulate directly by changing its excitation?
   a. Semiconductor
   b. Ruby
   c. Helium neon
   d. Neodymium YAG
   e. All equally difficult

9. What is the length of a Q-switched pulse from a 10-cm long neodymium-glass laser with a 90% reflective output mirror. Assume the refractive index of glass is 1.5.
   a. 1 nanosecond
   b. 5 nanoseconds
   c. 10 nanoseconds
   d. 20 nanoseconds
   e. 1 microsecond

10. A laser has a bandwidth of 4 gigahertz (4×10^9 Hz). What is the shortest modelocked pulse it can generate, according to the transform limit?
   a. 1 picosecond
   b. 10 picoseconds
   c. 30 picoseconds
   d. 100 picoseconds
   e. 250 picoseconds

Gas Lasers

ABOUT THIS CHAPTER

Gas lasers are one of the three major laser families, and are in many ways the most varied. In this chapter, we will learn how the basic concepts of laser operation apply to gas lasers. Then we will learn about the most important gas lasers.

THE GAS LASER FAMILY

The family of gas lasers is large and varied. The first gas laser was demonstrated at Bell Telephone Laboratories by Ali Javan, William R. Bennett Jr., and Donald R. Herriott in late 1960, only seven months after Maiman's first laser (which was ruby, a crystalline solid-state laser). Since then, laser action has been demonstrated at literally thousands of wavelengths in a wide variety of gases, including metal vapors, rare gases, and complex molecules.

Gas lasers have a wide variety of characteristics. Some emit feeble powers below a thousandth of a watt, but other commercial gas lasers emit thousands of watts. The most powerful continuous beams—a couple of million watts—have been generated by experimental military gas laser weapons. Some gas lasers can emit continuous beams for years; others emit pulses lasting a few billionths of a second. Their outputs range from deep in the vacuum ultraviolet—at wavelengths so short they are blocked completely by air—through the visible and infrared to the borderland of millimeter waves and microwaves.

What makes the gas-laser family so large? Researchers say it is partly the ease of testing different gases for laser action in the same tube. They pump out the old gas, pump in the new, close off the tube, and excite the gas. Such experiments are simple enough that some modest university laboratories ran up impressive lists of discoveries in the 1960s. The ease of experiments led to some accidental discoveries, such as the 488-nanometer line of the argon-ion laser. William Bridges at Hughes Research Laboratories in Malibu, California found that wavelength after putting argon into a tube and pumping it out while studying a mercury laser. Enough argon remained in the tube to generate visible light at 488 nm.

Today, semiconductor and solid-state lasers have begun replacing gas lasers for some applications, but gas lasers continue to play important roles in others. Semiconductor lasers are slowly creeping into the red end of the spectrum, but gas lasers remain the most common sources of shorter wavelengths. Gas lasers also offer the highest powers available in most parts of the spectrum. Although far more semiconductor lasers are sold than any other type, gas lasers account for a larger dollar volume than any other type—nearly $300 million in 1987, according to *Lasers & Optronics* magazine.

GAS LASER BASICS

Although gas lasers have many important differences, they also share some common features. *Figure 6-1* shows a generic gas laser. The laser gas is contained in a tube with cavity mirrors at each end, one totally reflecting and one transmitting some light to form the output beam. Most gas lasers are excited by passing an electric current through the gas; the discharge usually runs the length of the tube, as shown in *Figure 6-1*. Electrons in the discharge transfer energy to atoms or molecules in the laser gas, typically through at least one intermediate step. Then the excited gas emits light which resonates within the laser cavity and emerges to form the laser beam.

This picture is a very general one, and—as we will see later in this chapter—each gas laser has its own functional characteristics. However, the picture is a useful one, and it will help to elaborate on it before describing individual types of gas lasers.

Gas Laser Media

Many different gases are used in lasers. In developing the helium-neon laser, Javan's group at Bell Labs systematically studied energy levels in neon and other gases. During the great laser boom of the 1960s, some groups simply put different gases into tubes to see what worked. Now that laser technology is more mature, new gas

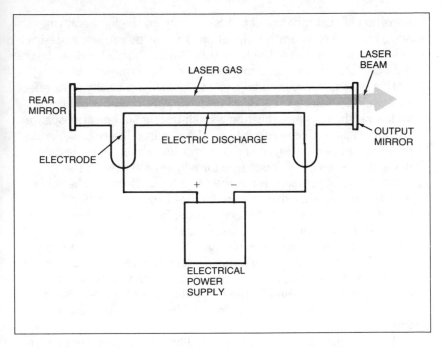

Figure 6-1. Generic gas laser.

lasers still are being discovered. Also, some old types are being developed commercially for the first time, but development has once again become systematic.

There are both obvious and subtle requirements for laser gases. The most obvious is that they must have sets of energy levels suitable for laser action. However, energy levels alone do not suffice. The laser medium must be in the right condition to populate those energy levels. Most practical gas-laser media are not pure gases, but mixtures of gases which serve different functions. For example, carbon-dioxide (CO_2) gas lasers contain helium and nitrogen as well as carbon dioxide. The nitrogen molecules absorb energy from the electric discharge and transfer it to CO_2 molecules. The helium atoms help CO_2 molecules drop from the lower laser level (maintaining the population inversion), and assist in heat transfer. In other lasers, extra gases are added to help absorb energy, transfer heat, or deactivate the lower laser level.

In practice, each laser has an optimum gas mixture. The

composition of that mixture depends not only on the light-emitting species, but also on operating conditions such as power levels, desired wavelength, and design of the laser tube itself. Thus, all carbon-dioxide lasers don't contain the same proportions of the different gases.

Gas pressure also is an important variable, especially in its effect on how well the laser gas conducts electricity. For most continuous-wave lasers, the pressure must be a small fraction of one atmosphere to sustain a stable electric discharge. Many pulsed lasers can operate at much higher pressures, sometimes over one atmosphere, because they do not need a stable discharge for a long time. Again, the optimum value of pressure is not the same for all lasers of the same type; the best pressure depends on details of the laser design.

Many gas lasers contain gases present in the atmosphere. Carbon-dioxide and helium-neon lasers are the most common examples, but the less-common nitrogen laser is even more striking because air—which is more than three-quarters nitrogen—can serve as the active medium. The gases are not always in the same form as in the atmosphere. For example, the 488 and 514.5 nanometer lines of argon are emitted by ions—atoms from which one electron has been stripped.

In other gas lasers, light is emitted by hot metal vapors which sometimes are ionized as well. Generally, the tubes of such lasers initially contain some solid metal that is partly vaporized to reach the pressure required for laser operation.

Gas Replacement, Flow, and Cooling

Many gas lasers normally operate with their tubes sealed, like a vacuum tube. However, some gas lasers require periodic fills of new laser gas, while in others the gas flows through the laser tube.

Early gas lasers needed periodic gas replacement because the tubes were not sealed well. Helium atoms, which are very small, can leak out of almost anything, and helium is an essential component of helium-neon lasers. Great strides have been made in glass-sealing technology, and helium-neon lasers now are rated to operate for 10,000 hours or more.

Some sealed gas lasers still require periodic replacement of the laser gas because contaminants accumulate, and gradually degrade laser action. This can be done in two ways, depending on how long the laser gas mixture normally lasts and whether or not other maintenance is needed when the gas is replaced. Excimer laser tubes normally are designed for periodic purging and refilling with fresh laser gas by the user. Argon-ion laser tubes are not designed for refilling, but the tubes can be sent back to a specialist that refurbishes them and fills them with fresh gas

In some higher-power gas lasers, the gas flows through the laser tube. This can serve the dual purpose of cooling the gas and removing contaminants. Many flowing-gas lasers operate in a closed cycle, but some lasers have an open cycle that requires removing the waste gas. Cooling systems also are needed for some sealed-tube lasers; if cooling requirements are modest, they may be met by a fan, but some high-power lasers require water cooling.

Excitation Methods

The most common way to excite gas-laser media is with an electric discharge passed along the length of the laser tube, as shown in *Figure 6-1*. The basic idea is similar to a fluorescent tube. First a high dc voltage is applied to "break down" the gas so it will conduct electricity. Once the gas becomes conductive, the voltage is reduced to a much lower level that will sustain the modest direct current needed to drive the laser.

This "longitudinal" excitation is fine for low-power lasers, but it cannot effectively deliver the high current needed for high-power pulsed or continuous-wave lasers. Such lasers usually are excited by a discharge applied perpendicular or "transverse" to the length of the laser tube.

Gas-laser power supplies must convert alternating current from the commercial power grid into different forms to drive lasers. For continuous-wave gas lasers, the power supply must raise the voltage and rectify the current to generate both the trigger voltage needed to break down the laser gas, and the steady voltage needed to operate the laser. Pulsed lasers require power supplies able to deliver very short high-energy pulses. Typically, such a power supply includes a dc module that charges a capacitor, and a fast discharge circuit that applies the electrical pulse across the laser, as shown in *Figure 6-2*. A key element is the switch that applies the high voltage across the laser gas. High-voltage, high-current switching technology is difficult, and the limitations of such switches typically limit the repetition rate of lasers with high pulse energy.

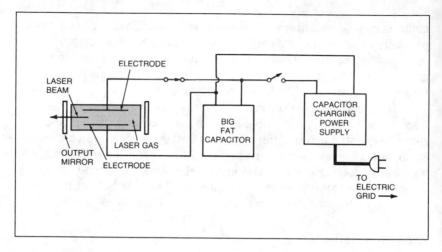

Figure 6-2. Pulsed laser power supply.

A few gas lasers are powered by zapping the laser medium with a beam of electrons from an accelerator. This technique is rare, and is used only to obtain very high pulsed powers.

As we will see at the end in this chapter, a few gas lasers are pumped optically. The usual source is an electrically excited gas laser with shorter wavelength, which raises the light-emitting atom or molecule above the upper laser level. Although the overall efficiency is limited, this approach can generate otherwise unobtainable wavelengths.

Tube and Resonator Types

If you go back and look at *Figure 6-1* carefully, you'll note it's a bit vague about placement of the mirrors. In practice, the mirrors can be combined with or separate from the windows at the ends of the laser tube. Combining the two can reduce the laser tube cost, but exposing the mirror coatings to the discharge in the laser tube can shorten their life. On the other hand, separating them increases costs, but improves performance.

The window at the end of the gas tube and the mirror at the end of the laser cavity have different functions. The window, like the rest of the laser tube, isolates the laser gas from air, preventing contamination and maintaining the required pressure. The window also should have loss as low as possible. The cavity mirrors also should have low loss, but their main function is to reflect laser light back and forth in the cavity to generate stimulated emission, and to couple some light out of the cavity as the laser beam.

One common design approach is shown in *Figure 6-3*. The windows at the ends of the laser tube are mounted at what is called Brewster's angle, which we described in the last chapter. As we indicated then, surface reflection depends on the angle from the normal, the refractive index, and polarization. If light strikes a transparent material with refractive index n from air, no light polarized parallel to the plane of incidence can be reflected if the light strikes at an angle θ_B given by:

$$\theta_B = \arctan n$$

For optical glass with a refractive index of 1.5, this Brewster's angle is about 57 degrees. Some light polarized perpendicular to the plane of the incidence is reflected, leading to higher losses for that polarization. Because of those excess losses, the laser beam is parallel to the plane of incidence of the Brewster-angle window.

In *Figure 6-3*, the cavity mirrors are curved, defining a confocal resonator. As we saw in Chapter 3, this stable-resonator design can provide a good-quality, diffraction-limited beam with low divergence.

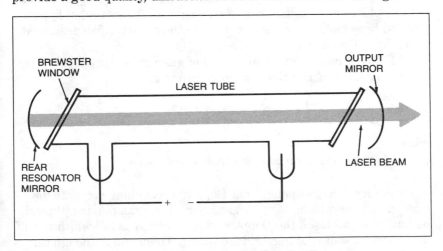

Figure 6-3. Laser tube with Brewster-angle window and confocal resonator.

This design is common in continuous-wave gas lasers at visible wavelengths, which typically have low gain and require optimized cavities to oscillate.

Different types of cavities are used in high-gain, pulsed lasers, notably the rare-gas halide excimer lasers described later in this chapter. Such lasers do not require as careful optimization to oscillate, and they use cavities where only the rear mirror is highly reflective. (The few percent reflection from an uncoated glass surface provides enough feedback to serve as an output mirror.) This design gives high-gain gas lasers that have comparatively large beam divergence and diameter.

One point to note is that such high-gain lasers can be operated in an oscillator-amplifier configuration. In such a configuration, a master oscillator generates a pulse which is amplified by passing it through a laser cavity without mirrors.

Wavelength and Bandwidth

Transitions in gas lasers, like those in other types, have a well-defined nominal wavelength. However, gas atoms and molecules emit at a broader range of wavelengths because they are continually moving at any temperature above absolute zero.

This effect is called Doppler broadening because it depends on the Doppler shift, the change in wavelength (or frequency) of light emitted by something moving relative to the observer. We described it briefly in Chapter 4.

The root mean square speed $\langle v \rangle$ of gas atoms is proportional to the atomic mass M and the gas temperature T:

$$\langle v \rangle = (3\,kT\,/\,M)^{\frac{1}{2}}$$

where,
k is the Boltzmann constant mentioned earlier.

This velocity is large enough that Doppler broadening accounts for most of the line widths of many gas lasers. For example, in a typical helium-neon gas laser, the Doppler width (defined as full-width at half maximum) is about 1.4 gigahertz, or about 0.0019 nm. Although this is small compared to the 4.738×10^{14} hertz frequency of the laser's 632.8-nm transition, it is much larger than the 1-megahertz bandwidth of one longitudinal mode of a typical helium-neon laser cavity. The Doppler width is broad enough to include several longitudinal modes, separated by about 500 MHz.

Doppler broadening occurs in gas lasers because the atoms are not fixed in place as they are in crystalline solid-state lasers or semiconductor lasers. However, other effects can broaden the range of wavelengths emitted by those types.

Differences among Gas Lasers

Although gas lasers share a number of common features, they also differ in many ways. Prominent among them are wavelength and output power, which are the most important laser characteristics for many applications. *Table 6-1* lists wavelengths and output power ranges for major commercial gas lasers; the most important types are described in more detail later.

Note that gas lasers can operate on electronic, vibrational, or rotational transitions. As we saw earlier, electronic transitions are at near-ultraviolet, visible, or near-infrared wavelengths. Vibrational transitions (which actually include rotational elements as well) are in the infrared, while rotational transitions are at the long-wave end of the infrared spectrum, and spill into the microwave region.

HELIUM-NEON LASERS

The helium-neon gas laser is the type best known to people outside the laser world. Often called the "He-Ne," its low-power red beam reads labels at supermarket checkout counters, aligns walls and surveying instruments at construction sites, and demonstrates how lasers work in countless school laboratories. Although semiconductor lasers now are more common because of their use in compact disc audio players, their presence is not obvious because they are deep inside the player and their beams are invisible near-infrared light. He-Ne lasers emit red beams which emerge into places you can see them, so the helium-neon laser is the type you are most likely to have encountered.

Physical Principles

The energy levels involved in the helium-neon laser are shown in *Figure 6-4*. Electrons passing through a mixture of five parts helium

Table 6-1. Major Gas Lasers, by Wavelength and Power Level, Grouped under Type of Transition Involved

Type	Wavelength (nm)	Power Range (W)‡ (Approximate)	Operation
Electronic Transitions			
Argon-fluoride excimer	193	0.5–50 (avg)	Pulsed
Krypton-fluoride excimer	249	1–100 (avg)	Pulsed
Xenon-chloride excimer	308	1–100 (avg)	Pulsed
Helium-cadmium (UV lines)	325	0.002–0.05	Continuous
Nitrogen	337	0.001–0.01 (avg)	Pulsed
Argon-ion (Ultraviolet lines)	350*	0.001–2	Continuous
Krypton-ion (UV lines)	350*	0.001–1	Continuous
Xenon-fluoride excimer	351	0.5–30 (avg)	Pulsed
Helium-cadmium	442	0.001–0.05	Continuous
Argon-ion	488–514.5	0.002–20	Continuous
Copper-vapor	510 and 578 nm	1–50 (avg)	Pulsed
Xenon-ion	540	—	Pulsed
Helium-neon	543	0.0001–0.001	Continuous
Gold-vapor	628	1–10	Pulsed
Helium-neon	632.8	0.0001–0.05	Continuous
Krypton-ion	647*	0.001–6	Continuous
Iodine	1300	—	Pulsed
Vibrational Transitions			
Hydrogen-fluoride (chemical)	2600–3000†	0.01–150	Pulsed or CW
Deuterium-fluoride (chemical)	3600–4000†	0.01–100	Pulsed or CW
Carbon-monoxide	5000–7000†	—	Pulsed or CW
Carbon-dioxide	9000–11000†	0.1–15,000	Pulsed or CW
Vibrational or Rotational Transitions			
Far-infrared	30,000–1,000,000†	⟨0.001–0.1	Pulsed or CW

*Other wavelengths also available
†Many lines in this wavelength range
‡For typical commercial lasers. Not indicated for lasers that are not often sold

to one part neon excite both species to high energy states, but the more abundant helium atoms collect more energy. As shown in *Figure 6-4*, the high-energy helium states have nearly the same energy as *5s* and *4s* energy levels of neon. (Those combinations of letters and numbers are the spectroscopic notations for electronic energy levels used in *Table 2-3*. Although they have specific meanings, it is much simpler just to regard them as labels. Some books use other notation.)

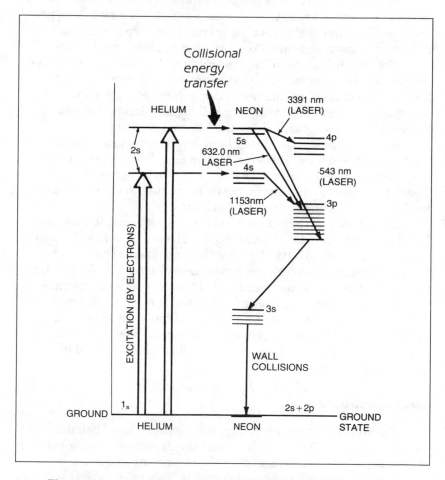

Figure 6-4. Energy levels and laser transitions in the helium-neon laser shown on a relative scale. (They actually are much higher above the ground state.)

The energy levels are close enough that the helium atoms can transfer excitation energy to neon atoms when they collide, which happens often in the gas. The $5s$ and $4s$ energy levels of neon are metastable, so atoms stay in those states for a comparatively long period of time.

Note that excitation raises neon atoms directly to a high energy level, without stopping at the intermediate levels shown at right in *Figure 6-4*. The result is a population inversion of neon, with more atoms in the upper levels than in some lower levels. Several energy levels are involved, and several laser transitions are possible in the helium-neon laser, depending on operating conditions and optics.

The first helium-neon laser operated at 1153 nm in the infrared, but the 632.8-nm red line was discovered soon afterwards, and has become the standard helium-neon laser wavelength. Recently, laser manufacturers have begun offering helium-neon lasers that emit at weaker visible lines, particularly the 543-nm green line.

After dropping to the lower laser level, neon atoms remain high above the ground level. However, they quickly lose that energy and drop through a series of lower energy levels to the ground state. From the ground state they can again be excited to the upper laser level by energy transfer from helium atoms.

Overall gain of the helium-neon laser is very low, so care must be taken to minimize laser cavity losses. The overall efficiency also is low—typically 0.01 to 0.1%—because the transitions are so far above the ground state. However, the helium-neon laser is simple, practical, and inexpensive; mass produced sealed-tube versions can operate continuously for tens of thousands of hours. It requires an ignition voltage of about 10,000 volts to break down the laser gas so it can carry an electrical current. After breakdown, a couple of thousand volts can maintain the current of a few milliamperes needed to sustain laser operation.

Laser Construction

The internal structure of a typical mass-produced helium-neon laser is shown in *Figure 6-5*. Note that the discharge passing between electrodes at opposite ends of the tube is concentrated in a narrow bore, one to a few millimeters in diameter. This raises laser excitation efficiency, and also helps control beam quality. The bulk of the tube volume is a gas reservoir containing extra helium and neon. Gas

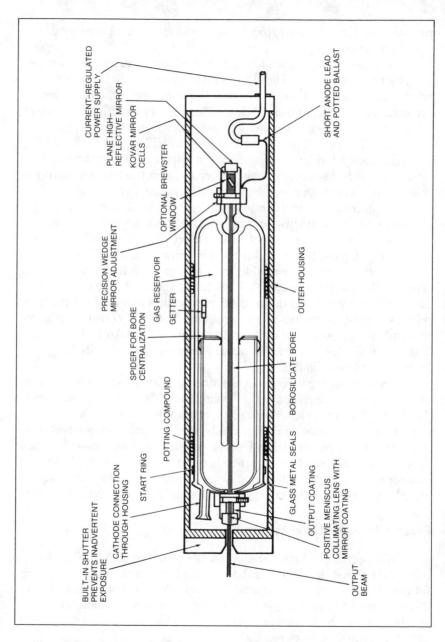

Figure 6-5. A mass-produced helium-neon laser. *(Courtesy Melles Griot)*

pressure within the tube typically is a few tenths of a percent of atmospheric pressure.

In these mass-produced helium-neon lasers, mirrors are bonded directly to the tube by a high-temperature process that produces what is called a "hard seal." This seal limits helium leakage, which otherwise might limit laser lifetime. The mirrors must have low loss because of the laser's low gain. The rear cavity mirror is totally reflective. The output mirror reflects most incident light back into the laser cavity, but lets a few percent escape in the laser beam. One or both mirrors have concave curvature to focus the beam within the laser cavity, which is important to good beam quality.

An alternative construction is to seal the laser cavity with Brewster's angle windows, and mount the mirrors separate from the laser tube. This approach is more expensive, but it avoids losses for the plane-polarized beam, and is used in some laboratory lasers.

The output power available from helium-neon lasers depends on tube length, gas pressure, and diameter of the discharge bore. Researchers have found that output power for a given tube length is highest when the product of gas pressure (in torr, 760 torr equals one atmosphere) times bore diameter (in millimeters) is 3.6 to 4. Extending the tube can raise output power somewhat, but the improvements are limited.

One interesting variant on the standard linear helium-neon laser is the ring laser, which has been developed for use as a rotation sensor. The tube for a "ring" laser actually is a triangle or square, with mirrors at each corner to reflect the beam from one arm into the next as it travels around the circumference. A ring laser can detect rotation about an axis perpendicular to the ring plane by measuring phase differences in light travelling in different directions around the ring. Ring-laser "gyroscopes" are used in some military and civilian aircraft.

Practical Helium-Neon Lasers

Mass-produced helium-neon lasers can deliver 0.5 to 10 milliwatts of red light. They range in size from some little bigger then fat pens, 10 centimeters long and 1.6 cm in diameter (4 by ⅝ inch) to 30 cm (a foot) or more long and 2–5 cm (1–2 inches) in diameter. A few special-purpose helium-neon lasers are considerably larger and can produce up to 60 mW of red light.

Although other visible wavelengths have been available

commercially since the mid-1980s, most people who mention helium-neon lasers assume they emit the 632.8-nm red line. Output at other visible wavelengths is much weaker, typically no more than a milliwatt. The most useful of the other wavelengths is the 543-nm green line, but 594- and 612-nm lines also are available. Powers of 1 to 10 mW are available on three infrared lines, at 1.153, 1.523, and 3.39 micrometers. Although a few models can be made to emit at different wavelengths by switching their optics, most helium-neon lasers are designed to emit only a single wavelength.

Mass-produced red helium-neon laser tubes may sell for $20 or less in large quantity, without power supply or accessories including safety equipment required by Federal regulations. Complete general-purpose helium-neon lasers ready for laboratory or other use start at a few hundred dollars each when bought singly. At that price, they remain the least-costly visible laser, although near-infrared semiconductor lasers do cost less. Prices are higher for helium-neon lasers that emit wavelengths other than the standard red line.

For most practical purposes, you can consider the light from a red helium-neon laser to be a single wavelength. The typical Doppler broadened bandwidth of a red helium-neon laser is 1.4 GHz, which corresponds to a coherence length of 20 to 30 centimeters, adequate for holography of small objects. If you need a longer coherence length or a narrower bandwidth, you must buy a special helium-neon laser which is limited to emit in a single longitudinal mode. The 1-MHz bandwidth of a typical single longitudinal mode corresponds to a coherence length of 200 to 300 meters.

Helium-neon lasers typically emit TEM_{00} beams, with diameter about a millimeter and diverence about a milliradian. You can expect a helium-neon laser tube with hard-sealed mirrors to operate for 20,000 hours or more. Because of their low cost, compact size, durability, and ease of operation, helium-neon lasers are by far the most widely used visible lasers.

RARE-GAS ION LASERS

The argon- and krypton-ion lasers bear enough resemblance to helium-neon lasers to be considered cousins. All are driven by electric discharges passing through elements of the rare-gas (Group VIII) column of the periodic table. All normally emit continuous beams of visible light.

However, there are many important differences. Argon- and krypton-ion lasers can generate much more power than helium-neon lasers, and produce the shorter wavelengths that are needed for some applications. Typically their output powers range from a few milliwatts to up to 20 watts. On the other hand, they are more delicate, much less efficient and much more costly. Most of their uses are specialized, but you may have encountered one: they are the preferred source for laser light shows.

One important note before we look inside rare-gas ion lasers: they are often called simply "ion" lasers. This isn't a very good choice of words because there are other gas lasers (such as helium-cadmium) in which the light is emitted by an ion, but it is a common one in the laser world.

Physical Properties

The active medium in rare-gas ion lasers is either argon or krypton at a pressure of roughly 0.001 atmosphere. (The two gases can be mixed to get emission on the wavelengths of both, but normally this is done only in lasers used for light shows.) Both gases emit light at several wavelengths spaced through the near-ultraviolet, visible, and near-infrared parts of the spectrum, listed in *Table 6-2*.

Unlike helium-neon lasers, the emission in argon and krypton lasers comes from atoms which have been ionized; that is, with one or two electrons stripped from their outer shells. Wavelengths shorter than 400 nm come from atoms with two electrons removed (Ar^{+2} or Kr^{+2}). Longer wavelengths come from singly ionized atoms (Ar^+ or Kr^+). Argon is a much more efficient laser gas, but krypton offers more wavelengths.

The internal kinetics of ion lasers are particularly complicated, but we can give an overview of what happens. As in helium-neon lasers, an initial high-voltage pulse breaks down the gas so it conducts current. Electrons passing through the gas transfer energy directly to the argon or krypton atoms, removing electrons, and raising the resulting ions to a group of high energy levels, as shown in *Figure 6-6*. Three processes can populate the metastable upper laser levels, from which the ions drop down to a group of lower laser levels. Argon and krypton lasers emit so many different laser lines because transitions can occur between many pairs of upper and lower levels; there isn't room to show all those levels in the diagram. If the optics

Table 6-2. Major Wavelengths of Ion Lasers

Argon	Krypton
334.0 nm	337.4 nm
351.1 nm	350.7 nm
363.8 nm	356.4 nm
457.9 nm	406.7 nm
476.5 nm	413.1 nm
488.0 nm (Strong)	415.4 nm
496.5 nm	468.0 nm
501.7 nm	476.2 nm
514.5 nm (Strong)	482.5 nm
528.7 nm	520.8 nm
1090.0 nm	530.9 nm
	568.2 nm
	647.1 nm (Strong)
	676.4 nm
	752.5 nm
	799.3 nm

permit it, argon and krypton lasers can oscillate simultaneously on several different visible wavelengths, each produced by a transition between a different pair of levels.

The lower laser levels all have a very short lifetime. Argon ions quickly drop from the lower laser level (an excited state of the ion) to the ion ground state by emitting extreme-ultraviolet light at 74 nm. The ground-state ion then can recapture an electron or again be excited to the upper laser levels. Similar things happen in krypton.

It takes much more energy to excite argon and krypton atoms to their laser levels than it does to excite neon atoms. In the helium-neon laser, it is atoms that emit light; in visible-wavelength argon and krypton lasers it is ions missing a single electron. Discharge currents in argon and krypton lasers are 10 to 70 amperes, more than a

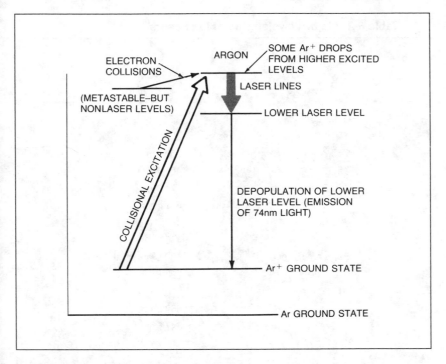

Figure 6-6. Energy levels that produce visible argon lines.

thousand times the level in a helium-neon laser, although the operating voltages of 90 to 400 volts are much lower. This high discharge current heats the laser gas to very high temperatures. (The ultraviolet lines of argon and krypton are emitted by ions missing two electrons, Ar^{+2} and Kr^{+2}. Those states lie much higher above the ground level than the singly ionized states that produce the visible lines, and thus require even higher current densities.)

The spectral line width of a single ion-laser line is about five gigahertz, broader than that of a helium-neon laser. This line width is large enough to allow modelocking, which generates pulses 90 to 200 picoseconds long. The main use of modelocked pulses is to pump the tunable dye lasers described in Chapter 9.

Argon- and Krypton-Ion Laser Structure

The structures of argon and krypton lasers are quite similar, and manufacturers often use the same basic laser tube for both types.

The difference is in the gas fill. A typical argon-ion laser may resemble a helium-neon laser at first glance, although ion lasers tend to be fatter and—especially at higher powers—larger. There also are some more subtle but nonetheless important differences.

Most argon and krypton lasers have Brewster-angle windows on the ends of the tube, with external mirrors defining the laser cavity. As in helium-neon lasers, the gain is low, so care must be taken to minimize losses within the laser cavity. That includes avoiding materials which suffer increased losses after being exposed to the extreme-ultraviolet light from decay of the lower laser level in argon. The tube itself usually is a ceramic. Cavity dumpers and modelockers can be inserted in the laser cavity to make the normally continuous-wave laser produce short pulses.

The laser cavity optics select which wavelengths argon and krypton lasers emit. If the goal is raw output power, the optics can allow oscillation at several wavelengths. If a single wavelength is needed, wavelength-selective optics can be inserted between the window and rear mirror to limit oscillation to a single laser wavelength. Standard cavity optics produce TEM_{00} beams.

As in helium-neon lasers, the discharge is confined to a narrow region in the center of the tube to enhance excitation efficiency. Older argon lasers confined the discharge in a narrow bore like that of the helium-neon laser, but in many newer lasers the discharge is confined by a series of metal disks with central holes that define a "bore." In either case, it is essential to provide a return path for the positive ions, which are attracted by the negatively charged cathode. A large gas reservoir also is needed because laser operation tends to deplete the gas.

With much higher output powers than helium-neon lasers but lower wall-plug efficiency (0.01 to 0.001%), argon and krypton lasers need some form of active cooling. Forced-air cooling is adequate for argon lasers delivering up to a few watts. At higher power levels, or for the less-efficient krypton laser and ultraviolet lines, water cooling is needed. Typically, tap water flows through pipes in the laser, then down the drain. (The need for both cooling water and high-voltage service limits where you can put high-power argon- or krypton-ion lasers.)

Practical Argon- and Krypton-Ion Lasers

As we indicated earlier, argon and krypton ions can oscillate on one or many wavelengths, depending on the cavity optics. If bright emission is required on lines throughout the visible spectrum, the two gases can be put into a single tube and excited simultaneously. Such mixed-gas lasers are sometimes used in laser light shows, but their lifetime is limited because krypton is depleted faster than the argon.

Most rare-gas ion lasers emit continuous wave TEM_{00} beams that are vertically polarized, but a few multiwavelength lasers may emit multiple modes. Although argon and krypton lasers have much higher-power output than helium-neon lasers, beam diameters and divergences are similar.

The prices of argon lasers have been coming down, but they remain much higher than those of helium-neon lasers. You can expect to pay thousands of dollars for the least expensive argon lasers, which emit several milliwatts. At the high end of the price scale, high-power argon lasers run tens of thousands of dollars. Krypton lasers cost more per watt because they are less efficient than argon.

Operating conditions inside argon and krypton laser tubes are extreme, so lifetimes are shorter than those of helium-neon lasers. Typical rated lifetimes are 1000 to 10,000 hours, with the longer lifetimes for lower-power tubes. The tubes account for much of the laser cost, and they are economically feasible to overhaul and refill.

METAL-VAPOR LASERS

There are two important families of commercial lasers in which light is emitted by a metal vapor. In one, typified by the helium-cadmium laser, the light emitter is an ionized metal. In the other, typified by copper- and gold-vapor lasers, light is emitted by neutral atoms. The distinction is important because their operating characteristics differ greatly.

Helium-Cadmium and Other Metal-Ion Lasers

Helium-cadmium (He-Cd) lasers produce continuous output at slightly higher power levels than helium-neon lasers, from under a milliwatt to tens of milliwatts. He-Cd wavelengths are shorter, 325 nm in the ultraviolet and a stronger 441.6-nm blue line. However, despite the similarities in name, power level, and some internal characteristics, the He-Ne and He-Cd lasers are not as closely related as they might at first seem. You can consider the helium-cadmium laser as in between argon and He-Ne lasers, in such characteristics as output power and overall efficiency.

The energy-level structures of helium and cadmium are shown in *Figure 6-7*. The laser gets its energy from electric current flowing through a thin capillary bore in the laser tube. The electrons excite helium atoms to high-lying states. Then the excited helium atoms transfer energy to cadmium atoms, ionizing them while raising them to the upper laser level. (Cadmium normally has only two electrons in an incompletely filled outer shell, and it is much easier to ionize than helium, argon, or krypton, all of which normally have full outer

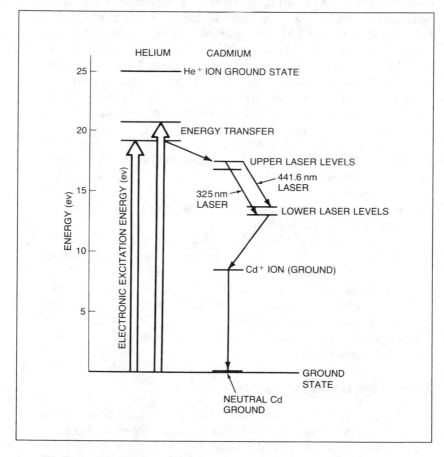

Figure 6-7. Energy levels in the helium-cadmium laser.

electron shells.) Cadmium atoms can be trapped in two metastable states, one of which produces the 325-nm line, the other the 441.6 nm line.

Because cadmium is a solid at room temperature, the metal in the laser tube must be heated to about 250° Celsius to produce the several millitorrs pressure of cadmium vapor needed for the laser to operate. Helium pressure is about a thousand times higher, several torrs, but still only about one percent of atmospheric pressure.

Like argon and He-Ne lasers, He-Cd laser tubes concentrate the discharge in a narrow bore to excite gas atoms efficiently. Chunks of cadmium metal are placed in the tube to replace cadmium that condenses on cool parts of the tube during laser operation. Tubes normally also include a helium reservoir to replace gas that leaks out. Together with zones of the tube designed to collect surplus cadmium so it does not deposit on critical optical surfaces, these reservoirs make He-Cd tubes look different from He-Ne and argon lasers. However, the basic technology is similar, except for the need to heat the metal. Helium-cadmium lasers need discharge voltages around 1500 V, and lower currents than argon lasers.

Optical design of helium-cadmium lasers also resembles those of He-Ne and argon lasers. Some He-Cd laser tubes have Brewster-angle windows and external resonator optics; others have cavity mirrors bonded directly to the tube. The output wavelength depends on the optics. Beam diameter and divergence are around one millimeter and one-milliradian.

In many practical terms, the He-Cd laser can be seen as an intermediate step between the He-Ne and the argon laser, although the wavelength is shorter than either. Lifetimes of blue He-Cd lasers are several thousand hours, comparable to those of argon lasers. Prices range from a few thousand dollars for a low-power model to over $10,000 for the most powerful types. Some similar metal-vapor ion lasers have been demonstrated in the laboratory, but none are available commercially.

Copper- and Gold-Vapor Lasers

Neutral metal-vapor atoms also are the basis of a family of gas lasers. The most important members of the family are the copper-vapor laser, which emits at 511 nanometers in the green and 578 nm in the yellow, and the gold-vapor laser, with a 628-nm red line. Unlike the other gas lasers we have described so far, the copper and gold vapor lasers are limited to pulsed operation. They can generate average powers up to tens of watts and thousands of pulses per second, but not a continuous beam.

The restriction to pulsed operation stems from the energy level structure of neutral metal vapors. When an electric discharge passes through copper vapor, collisions between electrons and copper atoms raise the copper atoms to one of two excited states. If vapor pressures are low, those states stay excited only about 10 nanoseconds, not long enough to produce laser pulses. However, if the density is increased (to pressures still well below atmospheric pressure), the states' effective lifetimes increase to 10 milliseconds, long enough for stimulated emission and laser operation. This quickly shifts the population to the two lower laser levels, which are metastable, having lifetimes of tens to hundreds of microseconds. The accumulation of atoms in the lower laser level stops laser action in less than 100 nanoseconds. Then the lower-level population decays, and soon the laser is ready to generate another pulse—at repetition rates of many thousands of pulses per second. Similar processes occur in gold and several other metals, producing lasers that can generate high average powers in rapid repetitive pulses.

To obtain the 0.1-torr vapor pressure needed for laser action, metallic copper or gold must be heated to 1500° to 1850° Celsius in the laser tube. This takes about half an hour, an exceptionally long warm-up time by laser standards. The metal vapor atoms are excited by collisions with electrons in a pulsed electric discharge passing the length of the tube. Addition of a rare gas (neon, argon, or helium) can improve discharge quality and speed depopulation of the lower laser level. During operation, the laser generates enough waste heat to keep metal vapor pressure high enough for laser operation. (In fact, metal-vapor lasers normally require active cooling with flowing water or forced air.)

Unlike the He-Ne, argon, and He-Cd lasers, copper-vapor lasers have high gain, 10 to 30% per centimeter. A copper vapor laser can operate without resonator mirrors, but commercial lasers have a totally reflective rear mirror and an output mirror that transmits about 90% of incident light, reflecting only about 10% back into the laser cavity. The laser tube windows are separate from the cavity mirrors, and the surface of uncoated window can reflect enough light back to serve as an output mirror.

Two factors control pulsing of gold and copper vapor lasers. The pulse length depends on internal kinetics, and how fast the lower laser fills up to stop the laser pulse, typically tens of nanoseconds. The repetition rate depends on the discharge electronics; each new light pulse must be triggered by an electrical pulse. Commercial copper and gold vapor lasers have repetition rates of several thousand hertz, which the human eye sees as a steady beam. With efficiencies of several tenths of a percent, copper and gold vapor lasers are the most efficient visible gas lasers and among the highest powered. Their repetitively pulsed output is not as useful as a continuous beam for many applications, and at tens of thousands of dollars, copper- and gold-vapor lasers are expensive. (Despite what you might think, the cost of gold adds little to the cost of a gold vapor laser. The laser consumes only about $20 worth of gold in an eight-hour day, and the gold is deposited in the tube where it can be recovered later).

CARBON-DIOXIDE LASERS

The carbon-dioxide (CO_2) laser is the most versatile gas laser, able to operate in either pulsed or continuous mode, and able to produce the highest continuous-wave power of any laser you can buy. Unlike argon, He-Ne, and metal-vapor lasers, the carbon-dioxide laser operates on a set of vibrational-rotational transitions. This puts its output at much longer wavelengths, 9 to 11 micrometers in the infrared region. There are several important types of carbon-dioxide lasers, but all work on the same transitions.

Basic Physics of CO_2 Lasers

The carbon-dioxide laser lines are emitted during transitions among three vibrational modes of the molecule shown in *Figure 6-8*, the symmetric stretching mode ν_1, the bending mode ν_2 and the asymmetric stretching mode ν_3. Each mode is quantized, so the molecule can have 0, 1, 2, etc. units of vibrational energy in each mode. In *Figure 6-8* the numbers in parentheses identify the energy levels in each vibrational mode. Rotational sublevels are not shown.

The laser transitions occur when CO_2 molecules drop from the higher-energy asymmetric stretching mode to the lower-energy symmetric stretching or bending modes. The transition to the symmetric stretching mode corresponds to a 10.5-micrometer photon;

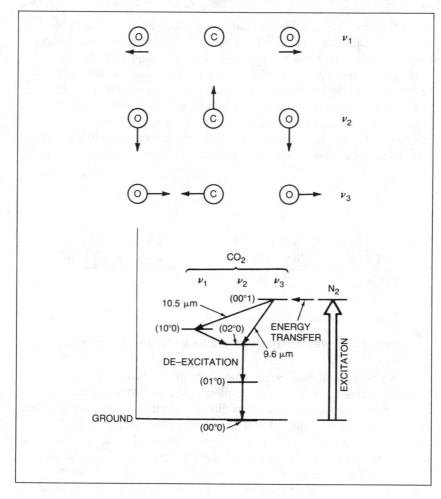

Figure 6-8. Vibrational modes and transitions of the CO_2 molecule.

dropping to the bending mode (actually, to the second excited level of the bending mode) corresponds to a 9.6-μm photon. However, the laser does not emit those precise wavelengths because the molecule changes its rotational state when it changes its vibrational state. The rotational transition energy is smaller than both thermal energy and the vibrational transition energy, so the change can be either up or down. That is, the molecule could speed up or slow down its rotation when moving between vibrational levels. Speeding up its rotation would consume some energy from the vibrational transition, so the emitted wavelength would be longer than the nominal transition energy. (For example, it might be 10.8 micrometers rather than 10.5 micrometers.) On the other hand, if the molecule slowed its rotation, the energy released would add to the energy from the vibrational transition, resulting in a shorter wavelength, such as 10.3 micrometers. Because of this phenomenon, carbon-dioxide lasers can emit a family of closely spaced wavelengths, as shown in *Figure 6-9*.

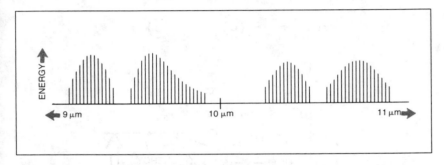

Figure 6-9. Wavelengths emitted by a carbon-dioxide laser.

An electric discharge passing through the laser gas excites a carbon-dioxide laser. The laser medium contains nitrogen and (usually) helium as well as carbon dioxide. Each gas serves a specific role. Both nitrogen and CO_2 absorb energy from electrons in the discharge. The lowest vibrational energy level of N_2 has nearly as much energy as the CO_2 molecule's asymmetric stretching mode, so N_2 can readily transfer energy to CO_2. The asymmetric stretching mode is the upper laser level of both groups of transitions. Helium helps to maintain the population inversion by getting CO_2 molecules to drop from the lower laser levels to the ground state or a lower bending level.

The carbon-dioxide laser is an exceptionally versatile system that can operate under a wide range of conditions. (There have even been reports that CO_2 laser lines are emitted by the upper atmosphere of Mars!) Its efficiency is exceptionally high by laser standards, and can reach up to 20%. It can produce a steady beam at low gas pressures, or powerful pulses at high pressures. It can operate simultaneously on many lines, or be tuned to a single wavelength. Commercial models routinely generate steady beams ranging from a few watts to many kilowatts, We will discuss the major types of CO_2 lasers in the following.

Types of CO_2 Lasers

Sealed-Tube Lasers

The simplest type of CO_2 laser operates in a sealed tube like the other gas lasers described earlier. As in those lasers, an electric discharge is applied along the length of the tube, with the positive electrode at one end and the negative electrode at the other end. Because the discharge breaks down CO_2 molecules to form oxygen and carbon monoxide, water or a catalyst must be added to the gas to regenerate CO_2, but this is not a major problem.

The sealed-tube design is convenient, and is widely used for lasers with powers under about 100 watts, priced at a few thousand to a few tens of thousands of dollars. However, other designs are required for higher power lasers. One reason is that the maximum output power depends on tube length; for sealed-tube lasers, a rule of thumb has been 50 watts per meter of gain medium. Mirrors can bend the laser beam inside the cavity so a 100-watt laser tube doesn't have to measure a full two meters end to end, but nonetheless, long tubes do become cumbersome. An added problem at high power levels is the need to remove waste heat.

Note that some gas flow is possible in a sealed-tube laser, by moving gas through the tube and the rest of a sealed system that includes a gas reservoir.

Waveguide CO_2 Lasers

A popular way to make compact carbon-dioxide lasers with powers from under a watt to about 50 watts is to shrink the tube's cross-section until it becomes a "waveguide." This is possible because the CO_2 laser's 10-micrometer wavelength is much longer than visible wavelengths. A waveguide laser is made like a dielectric waveguide used for microwave transmission, with internal dimensions of a millimeter or two across. These dimensions are small enough that the tube functions like a waveguide, avoiding large diffraction losses that otherwise would occur with an output aperture that is so small compared to the wavelength.

Gas must flow through the waveguide for such lasers to operate, but the waveguide can be made part of a sealed system with an internal gas reservoir. Low-power waveguide lasers can be slightly less expensive than low-power conventional sealed carbon-dioxide lasers. Normally operation is continuous.

Longitudinal Flow CO_2 Lasers

One way to increase power from a CO_2 laser is to flow fresh gas along the length of the laser cavity, in the same direction that the discharge current is applied. This consumes gas, but pressures are low and the gases used in CO_2 lasers are not particularly hazardous or costly. In addition, you can recycle at least some of the gas that has passed through the laser by mixing it with fresh gas.

The output power available per unit length is comparable to or somewhat higher than sealed-tube CO_2 lasers, but the design is straightforward and this approach has been widely used in continuous-wave lasers delivering hundreds of watts. As high-power devices, flowing gas lasers normally cost tens of thousands of dollars and up.

Transverse Flow CO_2 Lasers

Output power in a continuous CO_2 beam can reach about 10 kilowatts per meter of tube length if the laser gas flows perpendicular or transverse to the axis of the laser cavity (the line between the mirrors). This lets gas flow through the laser cavity much faster than if the gas had to pass along the tube axis. The faster gas flow removes waste heat and contaminants. The electric discharge that drives the laser also is applied perpendicular to tube axis, so it goes through a shorter length of gas.

Like longitudinal flow lasers, the gas pressure is low and the output beam is continuous. Typically, the gas is recycled, with some fresh gas added. Transverse flow normally is used only in very high-power lasers, with outputs in the kilowatt range or above, and prices to match. (A diagram of one early model with 15-kW output labelled part of the flow loop as a "wind tunnel.")

Gas-Dynamic CO_2 Lasers

An electric discharge is not the only way to produce a population inversion in carbon dioxide. Rapid expansion of hot, high-pressure CO_2 (typically mixed with other gases) through nozzles into a near-vacuum also can produce a population inversion, because the expansion reduces gas temperature, but does not drop all the molecules to low energy levels. The gas flows transversely through the laser cavity, as in a transverse flow laser, but does not require electrical excitation.

Twenty years ago, the first "gas-dynamic" carbon-dioxide laser represented a breakthrough in high-power lasers—the first laser to reach the 100-kilowatt range. That breakthrough triggered efforts beginning around 1970 to develop high-power laser weapons. The Pentagon spent hundreds of millions of dollars learning that gas-dynamic lasers don't make very good weapons.

Transversely Excited (TEA) CO_2 Lasers

All the variations on the carbon-dioxide laser we have described so far operate at pressures well below one atmosphere and normally generate continuous beams (although they can be made to produce pulses). The reason for the low pressure is that continuous electric discharges are not stable at pressures above about a tenth of an atmosphere.

An alternative is to increase gas pressure to around one atmosphere and pass a pulsed electric discharge through the gas. This works best if the discharge is transverse to the laser axis. Such lasers are called *transversely excited atmospheric-pressure* or "TEA" lasers, although they do not always operate at a pressure of one atmosphere.

TEA lasers are compact sources of intense pulses lasting from 40 nanoseconds to a microsecond. They cover a wide range in power levels, size, and price—the latter ranging from a few thousand dollars to several tens of thousands.

Optics and Wavelength Selection

We mentioned earlier that carbon-dioxide lasers can produce a broad range of wavelengths because of the many combinations of rotational and vibrational transitions. The range of wavelengths has a significant impact on the design of laser optics.

The carbon-dioxide laser has good but not spectacular gain, so it normally operates with a totally reflective rear mirror and a partly reflective output mirror. Often the cavity mirrors are made of metal, with output coupling through a hole in the mirror rather than through a partly transmissive coating. Because conventional silica glasses are not transparent at 10 μm, the output windows of a CO_2 laser tube may not look transparent to you, but they are transparent to the laser beam.

Many carbon-dioxide lasers generate the entire range of possible wavelengths, because for applications such as materials working there isn't much difference between 9 and 11 μm. For drilling or cutting, the main concern is delivering the maximum power to the focal spot—which requires high power and a good-quality beam. These lasers are made with cavity optics reflective throughout the 9 to 11 μm range, to extract as much energy as possible from the CO_2 laser cavity.

On the other hand, scientific applications often require one specific CO_2 wavelength. Lasers for such applications are made with internal tuning optics that can limit oscillation to one line from the entire range of possible transitions. As long as the laser is operated at low pressure, you can select from many discrete lines. However, as pressure increases, the individual lines broaden. If TEA lasers are operated at pressures above about 10 atmospheres the separate rotational lines blend together to form a continuous spectrum throughout the CO_2 laser's operating range.

10-μm Quirks

The 10-μm part of the spectrum is a strange world to those of us used to visible light. As we mentioned earlier, glass is not transparent at 10 μm, but other materials are. Some optical materials are transparent both at visible and 10-μm wavelengths, including salt (sodium chloride) and zinc sulfide. However, others aren't, so you can't assume that something that looks opaque isn't emitting 10-μm light.

Infrared viewers can show you carbon-dioxide laser beams, but you need the right kind of infrared viewer. Many operate in the near infrared, at wavelengths near one micrometer, and those won't do. You need a "thermal" infrared viewer. The name comes from the fact that thermal radiation from room-temperature objects peaks near 10 μm, so everything around you will look bright at that wavelength. (The military uses thermal viewers to search for enemy soldiers and vehicles, both of which are hotter than their surroundings, and the military developed much of the infrared technology used with CO_2 lasers.)

One other note about carbon-dioxide laser optics: so far there are no good optical fibers for that wavelength. People have tried to make many materials transparent at 10 μm into optical fibers, but haven't had much success. If you have any bright ideas for 10-μm optical fibers, there are lots of people who would like to talk with you.

CARBON-MONOXIDE LASERS

The carbon-monoxide (CO) laser is something of a less-successful brother of the carbon-dioxide laser. Emitting on vibrational-rotational transitions lying mostly between 5 and 6 μm, it can be even more efficient than CO_2. Like CO_2, the CO laser can emit powerful continuous beams. It has been plagued by serious practical problems, including strong absorption of some lines by the atmosphere, and the need to cool the gas to well below room temperature for efficient operation. However, research and development continues, and some observers hope it will soon find medical applications.

EXCIMER LASERS

Excimer lasers are a family of lasers in which light is emitted by a short-lived molecule made up of one rare gas atom (e.g., argon, krypton, or xenon) and one halogen atom (e.g., fluorine, chlorine, or bromine). The most important excimer lasers are listed in *Table 6-3*. First demonstrated in the mid-1970s, excimer lasers have become important because they are the most powerful practical ultraviolet lasers. They also rely on an interesting and unusual set of physics.

Physical Fundamentals

Rare-gas halides are peculiar molecules that emit laser light on an unusual type of electronic transition. The two atoms are bound only when the molecule is in an excited state. That is the upper laser level. When the molecule drops to the ground state, which is the lower laser level, the molecule falls apart. That produces a population inversion in a rather unusual way—there can't be any molecules in the lower laser level because they are not bound together.

Figure 6-10 shows the energy levels of a typical rare-gas halide as a function of the spacing between the two atoms in the molecule, R (the rare gas) and H (the halide). The dip in the excited-state curve shows where the molecules are metastable; the absence of a dip in the

Table 6-3. Major Excimer Lasers

Type	Wavelength
F_2*	157 nm
ArF	193 nm
KrCl	222 nm
KrF	249 nm
XeCl	308 nm
XeF	350 nm
*Not a rare-gas halide, but usually grouped with excimers	

ground-state curve indicates that the molecules fall apart. When the molecule is excited, the energy is at a minimum when the two atoms are a certain distance apart, trapped in a potential well. When they are in that potential well, they can occupy several vibrational levels as well (shown as horizontal lines in the potential well). However, in the "ground state," with the lowest possible energy, there is no bonding energy to hold the two atoms together, and the molecule falls apart, as shown in the lower curve. This reflects something you may remember from elementary chemistry—rare gases don't like to form compounds, even with elements as highly reactive as halogens.

Excimer lasers are excited by passing a short, intense electrical pulse through a mixture of gases containing the desired rare gas and halogen. Normally, 90% or more of the mixture is a buffer rare gas (typically helium or neon) that does not take part in the reaction. The mixture also contains a small percent of the rare gas (argon, krypton, or xenon) that becomes part of the excimer molecule, and a smaller fraction of molecules that supply the needed halogen atoms. The halogen atoms may come from halogen molecules such as F_2, Cl_2, or Br_2, or from molecules that contain halogens such as nitrogen trifluoride (NF_3). The advantage of avoiding pure halogens is that they are very reactive. Fluorine, in particular, is so treacherous to handle that the developers of one high-energy laser that used fluorine spoke of "the fire of the week."

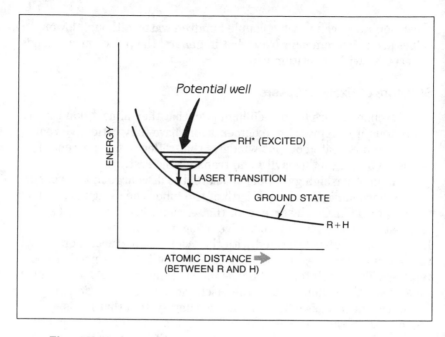

Figure 6-10. Internal energy of a rare-gas halide molecule in excited and ground states.

Electrons in the discharge transfer energy to the laser gas, breaking up halogen molecules and causing formation of electronically excited molecules like xenon fluoride (written XeF*, with the * meaning excited). The reactions involved are very complex, and depend on the type of gases. The molecules remain excited for about 10 nanoseconds, then drop to the ground state and dissociate. The molecular kinetics (as well as the duration of the driving electrical pulses) limit laser operation to pulses lasting tens or hundreds of nanoseconds. The energies involved are large, and output is at ultraviolet wavelengths.

Excimer-laser repetition rates depend more on the power supply than on the gas. The principal limitation is speed of the high-voltage switches. The highest repetition rates are around 1000 hertz, but more typical values are tens to a few hundreds of hertz. Pulse energies range from about 10 millijoules to a few joules, and differ somewhat among gases, with KrF and XeCl generally the most energetic. Average power—the product of pulse energy times repetition rate—can reach a couple hundred watts, although lower values are more common. Note that in general the pulse energy tends to decrease with repetition rate.

Structure of Excimer Lasers

Excimer lasers have such high gain that they almost don't need cavity mirrors. In practice, excimer lasers have fully reflective rear mirrors and uncoated output windows that reflect a few percent of the beam back into the cavity and transmit the rest.

As in other high-gain pulsed lasers, the discharge in an excimer laser is perpendicular to the length of the tube. The design is similar to that of a TEA CO_2 laser, except that excimer laser tubes must resist attack by the highly corrosive halogens in the laser gas. Excimer laser tubes are filled with the laser gas mixture, then sealed and operated for a certain number of shots until the gas needs to be replaced. The tube's total volume is much larger (typically 100 to 1000 times) than the volume where the discharge excites laser action. Often, the gas is passed through a recycling system that helps regenerate the proper gas mixture and extend the life of the gas fill. The laser's pulse energy drops with time, until the spent gas must be pumped out of the laser and replaced. The number of shots depends on the gas, and can be many millions of shots for longer-lived gases such as xenon chloride.

Although that number of shots may sound impressive, a little multiplication will show that at high repetition rates it doesn't amount to very much time. A 200-hertz laser generates $200 \times 60 \times 60$ pulses an hour—720,000 pulses! Thus, a gas supply is part of any excimer laser set-up.

Practical Excimer Lasers

Excimer laser technology is comparatively young, and most excimer lasers are used in the laboratory. However they are the best available pulsed ultraviolet laser, with wall-plug efficiency as high as a couple of percent. They are beginning to appear in medical research and in high-technology industrial systems, particularly for the manufacture of semiconductor electronics. As the handful of companies that make excimer lasers tries to satisfy industrial and medical users, their products will become more reliable, although the presence of halogens does present some formidable problems.

Laboratory excimer lasers have long been designed to handle any of several gas mixtures. This reflects the needs of a research laboratory, which one day may be working with the 308-nm xenon-chloride line, and the next day may need the 193-nm argon-fluoride wavelength. The researcher would pump out the old gas mixture, passivate the tube to remove contaminants, then pump out that mixture and replace it with a new laser gas mixture. This capability is not essential for industrial applications, and such lasers often are made for a specific gas mixture.

Excimer lasers remain complex systems and on the pricey side, typically in the tens of thousands of dollars, with the average price probably $35,000 to $50,000.

NITROGEN LASERS

The 337-nm nitrogen laser bears some resemblances to excimer lasers. Like excimer lasers, it has high gain and can produce short pulses with high peak powers. It operates on a combined electronic-vibrational transition of molecular nitrogen, which is excited by a pulsed electric discharge at pressures from about 0.03 to one atmosphere. The excitation mechanism is efficient, but other processes in the laser are not. The lower laser level has a 10-microsecond lifetime, long enough to terminate the population inversion and laser action quickly. As a result, laser efficiency is 0.1% or less, pulse length is limited to a few nanoseconds, and pulse energy is limited to about 10 millijoules.

Because the low pulse energy limits average power, many customers have turned to excimer lasers, which can provide much more power because of their higher pulse energy. However, unlike excimer lasers, nitrogen lasers can readily be made small and cheap. Nitrogen lasers also avoid the need to handle dangerous gases. Compact nitrogen lasers that generate 0.1-millijoule pulses now are available for $1000 to $2000, and are used in measurement and remote sensing systems.

CHEMICAL LASERS

One of the more interesting laser types is the chemical laser. The most common type operates on vibrational transitions of hydrogen fluoride (HF) in the infrared. The standard HF laser emits light on many lines between about 2.6 and 3.0 micrometers. If normal hydrogen-1 is replaced by the heavier isotope deuterium (hydrogen-2), the wavelength is shifted to 3.6 to 4.0 micrometers, and the laser is called a deuterium-fluoride (DF) laser.

It is possible to excite chemical lasers by passing an electric discharge through a mixture containing hydrogen and fluorine atoms (often in other molecules, such as sulfur hexafluoride, SF_6). Alternatively, the excitation can be purely chemical, caused by the combustion of hydrogen with fluorine.

The HF laser relies on a chemical chain reaction:

$$H_2 + F \rightarrow HF^* + H$$
$$H + F_2 \rightarrow HF^* + F$$
$$H_2 + F \rightarrow HF^* + H$$
$$H + F_2 \rightarrow HF^* + F$$
$$\text{ad infinitum}$$

Each step produces a vibrationally excited HF molecule which emits an infrared photon. The chain reaction can continue as long as there are enough molecules to supply hydrogen and fluorine atoms. The chemical laser can operate continuously as long as the gases are flowing rapidly through the laser and the laser cavity, as shown in *Figure 6-11*. Here, helium is a buffer gas; other gases may be added to control production of fluorine. Although the population inversion doesn't last very long in any particular group of gas molecules, the gas flows through the laser so fast that it doesn't matter—fresh gas moves into the laser cavity between the resonator optics, maintaining a steady population inversion that can produce a continuous beam.

The overall effect of a chemical laser is something like a rocket engine. Combustion of fuels containing hydrogen and fluorine generates energy that is released as a laser beam. The reaction consumes the fuels, which must be pumped out of the laser and collected. (Hydrogen fluoride is nasty stuff, too.) The engineering is tricky, but it can draw upon earlier work with rocket engines.

Some laboratory-scale hydrogen- and deuterium-fluoride lasers are made that can emit up to tens of watts. However, most interest in chemical lasers comes from military developers of high-energy laser weapons. One developmental deuterium-fluoride laser the size of a building is reported to have produced continuous powers of two million watts at the Army's White Sands Missile Range in New

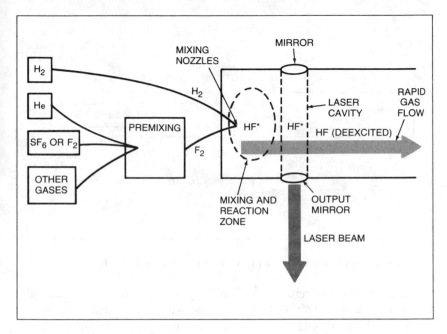

Figure 6-11. Basic operation of a chemical laser.

Mexico. However, operation of the laser—called MIRACL, for Mid-InfraRed Advanced Chemical Laser—is possible for only brief intervals and requires many laser specialists. Practical laser weapons are many years away, and require developments other than merely building monster lasers. (Key problems include optics to focus the beam, control of atmospheric effects, systems to identify targets and point the laser beam at them, and practical ways to put large and powerful lasers into space.)

FAR-INFRARED LASERS

Earlier we mentioned lasers that can emit infrared wavelengths much longer than the 10-micrometer output of the carbon-dioxide laser. These are the so-called "far-infrared" lasers, in which molecular gases emit at wavelengths between 30 and about 1000 millimeters.

As shown in *Figure 6-12*, the energy that powers a far-infrared laser comes from a shorter-wavelength infrared laser beam, typically one of the 10-μm lines of a CO_2 laser. A specific narrow wavelength

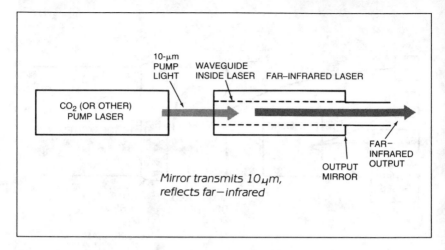

Figure 6-12. Optical pumping of a far-infrared laser.

range is needed to excite the molecular gas to an excited vibrational state. The laser transition takes place between two rotational levels in the excited vibrational state. The CO_2 laser beam enters the far-infrared laser cavity at one end, and excites molecules as it passes along the tube. The far-infrared beam exits at the other end. The end mirrors contain holes, in the back mirror to let the CO_2 beam in, and in the output mirror to let the far-infrared beam out.

Far-infrared lasers are unusual by laser standards because of the strange nature of far-infrared optics. They have few applications, but are an important variant on the gas laser.

What Have We Learned?

- Gas lasers range from weak to powerful. Their wide variety reflects the ease of developing gas lasers by testing gases in laser tubes.

- Most lasers contain mixtures of gases. For each gas laser, there is an optimum gas mixture and pressure.

- Gas lasers can operate with sealed tubes, but some require flowing gas.

- An electric discharge passing the length of the laser tube excites most low-power lasers. Higher-power lasers are excited with a discharge transverse to the tube.

- Pulsed lasers can be powered by charging a capacitor and discharging it through the gas.

- In many gas lasers, windows are mounted at Brewster's angle to minimize losses, and mirrors are separate from the tube. Brewster windows polarize the beam.

- Stable resonators are used for low-gain continuous-wave gas lasers. Other lasers with higher gain use different cavity designs.

- Thermal motion of gas atoms causes Doppler broadening, which can dominate a gas laser's bandwidth.

- The red helium-neon laser is the type you are most likely to encounter.

- An electric discharge excites helium atoms in a helium-neon laser; the helium atoms transfer energy to neon to produce a population inversion.

- The 632.8-nm red line is the most common from helium-neon. The 543-nm green line is weaker but still useful.

- Output from a He-Ne depends on tube length, gas pressure, and discharge bore diameter.

- Coherence length of a typical helium-neon laser is 20 to 30 cm, adequate for holography.

- Argon- and krypton-ion lasers can emit more visible power, and operate at shorter wavelengths, than the helium-neon laser.

They also emit several near ultraviolet and visible wavelengths, corresponding to different transitions.

- More energy is needed to excite argon and krypton to their upper laser levels, so they are less efficient than He-Ne lasers.
- Argon and krypton lasers have very similar structures, and somewhat resemble the helium-neon laser.
- Helium-cadmium lasers produce milliwatts continuous wave at 441.6 nm, and have a weaker line at 325 nm.
- Cadmium is a solid at room temperature, so the tube of a helium-cadmium laser must be heated to produce cadmium vapor.
- Neutral metal atoms emit fast, repetitive pulses in copper and gold vapor lasers.
- Copper vapor lasers have much higher gain than helium-neon, He-Cd, or argon lasers. They are among the highest powered visible lasers.
- The carbon-dioxide laser is the most versatile and highest-powered gas laser. It operates on vibrational transitions at 9 to 11 micrometers.
- CO_2 lasers make transitions among three vibrational modes of the CO_2 molecule. Rotational sublevels create many closely spaced lines.
- Carbon-dioxide lasers in sealed tubes can emit up to 100 W.
- CO_2 can oscillate in a waveguide structure a millimeter or two across, to generate laser powers to about 50 watts.
- Gas flows along the length of the tube in longitudinal flow CO_2 lasers, allowing higher power than sealed tube lasers.
- Gas flow perpendicular to the laser axis can raise CO_2 laser output to about 10 kW/meter of tube length.
- Rapid expansion of hot, high-pressure CO_2 produces a population inversion in a gas-dynamic laser.
- TEA CO_2 lasers produce high-power pulses when a transverse electric discharge is passed through the laser gas at a pressure near one atmosphere.
- CO_2 lasers can operate on one line or many lines, depending on the choice of optics.

- Carbon-monoxide lasers emit on vibrational-rotational transitions at 5 to 6 μm.

- Short-lived molecules containing one rare gas atom and one halogen atom emit intense ultraviolet pulses in excimer lasers. The molecules break up in the ground state, so the lower laser level always is empty.

- Excimer kinetics limit pulse lengths to tens or hundreds of nanoseconds; the repetition rate depends on the pulsing electronics.

- Excimer lasers have very high gain, so they need little feedback from the output mirror. One gas fill can generate millions of shots.

- Nitrogen lasers have high gain and emit 337-nm pulses with high peak power, but they have low average power and efficiency. They can be made compact and inexpensive.

- Chemical lasers get their energy from the reaction of hydrogen and fluorine to produce vibrationally excited hydrogen fluoride.

- Far infrared lasers operate on vibrational and rotational transitions at 30 to 1000 micrometers.

WHAT'S NEXT?

In Chapter 7 we will learn about solid-state crystalline and glass lasers, which have different properties than gas lasers. We will cover semiconductor lasers in Chapter 8, and dye and other lasers in Chapter 9.

6

Quiz for Chapter 6

1. Which of the following general statements is not true about gas lasers?
   a. Most gas lasers are excited electrically
   b. Only materials which are gaseous at room temperature can be used in gas lasers
   c. The laser tube cannot contain atoms or molecules other than the one species that emits light
   d. Gas lasers can emit continuous beams
   e. Gas must flow continually through a gas laser

2. What type of laser cavity should be used with a low-gain continuous-wave gas laser?
   a. Stable resonator, confocal
   b. Plane-parallel resonator
   c. Unstable resonator
   d. Any of the above
   e. None of the above

3. Which of the following lasers can emit the shortest wavelength in a continuous-wave beam?
   a. Helium-neon
   b. Helium-cadmium
   c. Nitrogen
   d. Argon-fluoride
   e. Carbon-dioxide

4. Which of the following lasers emits on a vibrational transition?
   a. Krypton-fluoride excimer
   b. Helium-neon
   c. Nitrogen
   d. Carbon-dioxide
   e. Krypton-ion

5. What excites the helium-neon laser?
   a. An electric discharge passing through the gas
   b. Transfer of energy from helium to neon
   c. Energy retained by the laser mirrors
   d. A and B
   e. None of the above

6. What kind of helium-neon laser can't you find commercially?
   a. A two-watt CW laser emitting at 3.39 μm
   b. A 0.5-milliwatt CW laser emitting at 543 nm
   c. A 20-milliwatt CW laser at 632.8 nm
   d. A 0.5-milliwatt CW laser emitting 632.8 nm
   e. You can buy anything if you have enough money

7. What gas laser has strong lines at 488 and 514.5 nm?
   a. Helium-neon
   b. Copper-vapor
   c. Helium-cadmium
   d. Carbon-dioxide
   e. Argon-ion

8. Which gas laser has the highest overall efficiency?
   a. Nitrogen
   b. Excimer
   c. Carbon-dioxide
   d. Helium-neon
   e. Argon-ion

9. Which of the following lasers cannot emit continuous wave because its internal kinetics automatically stop pulses?
   a. Excimer
   b. Nitrogen
   c. Copper vapor
   d. Gold vapor
   e. All of the above

10. Which of the following is not a type of carbon-dioxide laser?
    a. Gas dynamic
    b. Sealed tube
    c. Waveguide
    d. Optically pumped
    e. TEA

Solid-State Lasers

ABOUT THIS CHAPTER

In this chapter, you will learn about solid-state lasers, in which light is emitted by atoms in a crystal or glassy material. After first explaining the basic concepts of solid-state lasers, we will describe the most important types: ruby, neodymium, and vibronic lasers.

WHAT IS A SOLID-STATE LASER?

The laser world has come to use the term "solid-state" in a special sense, which we need to clarify right away. A solid-state laser is one in which the atoms that emit light are fixed within a crystal or a glassy material. It is not the same as a semiconductor laser, even though semiconductors are crystalline materials. In the laser world, semiconductor lasers belong in a separate category, described in the next chapter.

This may seem confusing if you're familiar with electronics. In electronic terminology, "solid-state" is synonymous with semiconductor. That usage dates from the days when transistors started replacing vacuum tubes, and the goal was to show that the new electronics relied on different principles than the old. Since semiconductor electronics was an outgrowth of solid-state physics, the label solid-state was attached to them.

The laser world makes a distinction between solid-state and semiconductor lasers for much the same reason the electronic world differentiates between vacuum-tube and semiconductor electronic devices—they rely on different principles and have different characteristics. Solid-state lasers are electrically nonconductive, and must be excited optically by light from an external source. Semiconductor lasers are often called diode or injection lasers because they rely on specific electronic properties of a junction

7

between two semiconductor materials, as you will learn in Chapter 8. Be sure to remember that difference.

PRINCIPLES OF SOLID-STATE LASERS

Theodore Maiman's first laser was a solid-state laser, ruby, as were the second and third lasers discovered. The operation of solid-state lasers has been refined greatly since then, but—as with gas lasers—the same basic principles underly the operation of the entire family of solid-state lasers, such as the generic type shown in *Figure 7-1.*

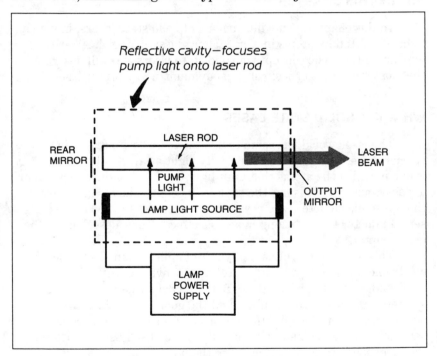

Figure 7-1. A generic solid-state laser.

The atoms that emit light in solid-state lasers are dispersed in a crystal or glass that contains many other elements. The crystal is shaped into a rod, with mirrors placed at each end. Light from an external source—pulsed flashlamp, a bright continuous arc lamp, or another laser—enters the laser rod and excites the light-emitting atoms. The cavity mirrors form a resonant cavity around the inverted population in the laser rod, providing the feedback needed to generate a laser beam that emerges through the output mirror. If the laser is pumped by a lamp source, like our example in *Figure 7-1*, the lamp and the laser rod are enclosed in a reflective cavity that focuses the pump light onto the rod.

Both the characteristics of the laser medium and of the pump light are important in determining how solid-state lasers operate. Let's first look at material characteristics, then at optical pumping techniques and sources.

Solid-State Laser Materials

We saw earlier that in many gas lasers only a small fraction of the atoms or molecules in the laser tube belonged to the species that actually emitted light. The same is true in all solid-state lasers. The atoms that emit light are embedded in a glass or crystalline matrix, as shown in a very simplified form in *Figure 7-2*. (Actual laser crystals have more complex structures.) From a chemical standpoint, the light-emitting element is a dopant added to a compound that serves as a crystalline host. Typically the light-emitting species accounts for somewhere around one percent of the material. (In some developmental laser materials, the light-emitting species is a minor part of the crystalline compound, but these are not in practical use.)

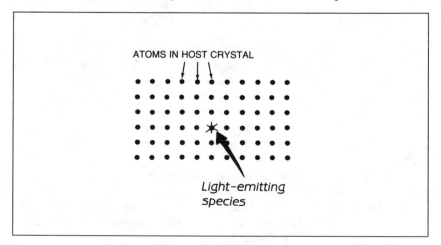

ATOMS IN HOST CRYSTAL

Light–emitting species

Figure 7-2. Light-emitting atoms in a solid-state laser are embedded in a crystalline or glass host.

7

The laser characteristics depend both on the light-emitting species and on the host.

The light-emitting species is the key ingredient in a solid-state laser. It must have a set of suitable energy levels that let it absorb pump light to populate a metastable upper laser level. (In some developmental solid-state lasers, one species absorbs the light energy and transfers it to another species in the crystal that emits the light, but such lasers have yet to emerge from the laboratory.) The best light emitters for solid-state lasers are a handful of related metallic elements: chromium, neodymium, erbium, holmium, cerium, cobalt, and titanium. Chromium is the light-emitting species in ruby lasers, and in commercial alexandrite lasers. The most common types of solid-state lasers rely on emission from neodymium.

We should mention that the light-emitting species often are called "ions" because of their nominal chemical valence in host crystals. The most common light emitters all have nominal ionic states of +3 e.g., Cr^{+3}, Nd^{+3}, Er^{+3}, and Ho^{+3}. You should consider those ionization states more nominal than real. Chemically, these elements are not completely ionized in the crystal, and they form bonds with atoms in the host material that are at least somewhat covalent. The "ions" are fixed in the crystal, and the "missing" electrons are nearby, typically forming a bond with oxygen atoms.

Host Materials

The interaction between light-emitting species and host material plays an important role in the operation of solid-state lasers. Potential host media must meet several requirements. The most obvious is that the host be reasonably transparent to the pump light, and absorb very little light at the laser wavelength. Too much absorption at either wavelength could greatly reduce efficiency, or make laser action impossible. They also could contribute to heating of the host, which can impair laser action.

The thermal properties of the host also are important. As in other lasers, most of the pump energy winds up as heat rather than laser light. Because atoms in laser crystals are fixed, they are not as efficient in removing heat as gas atoms. Other thermal problems also can arise, as we will learn later in this chapter.

The light-emitting species interact with the host crystal in subtle ways that influence its energy-level structure. Crystalline bonds and effects of adjacent atoms slightly shift energy levels in the light-emitting species. This can change the laser wavelength, usually by a small amount. For example, neodymium emits at 1054 nm when it is doped into phosphate-based glass, and at 1064 nm when in a crystalline host known as "YAG" (for yttrium aluminum garnet).

The basic transitions of the light-emitting atoms are electronic, arising from shifts of electrons between energy levels. When those atoms are embedded in crystals, their electronic energy levels can interact with vibrational energy levels in the crystal to produce what are called "vibronic" transitions, in which both electronic and vibrational energy levels change. Because vibrational transitions involve much less energy than electronic transitions, the main factor determining laser wavelength is the electronic energy levels of the light-emitting species. However, if there are many vibrational transitions possible, they can smear out the crystalline electronic transition over a range of wavelengths. As we will see later in this chapter, this gives some solid-state lasers gain over a wide enough bandwidth to tune their output wavelength.

Solid-State Laser Development

There are fewer types of solid-state lasers than there are of gas lasers. The major reason is not that solid-state media are less (or more) suitable for lasers, but the way that development must be conducted. To test a gas laser, you need only put the desired gas mixture into an evacuated tube. To test a solid-state laser, you have to grow a crystal containing the desired concentration of light-emitting dopant in a suitable host. Crystal growth is a complex art; it is no easy matter to grow an optically flawless sample suitable for manufacture into a laser rod. Nor is it easy to produce a rod from the crystal. This means that a given investment of time and money is likely to produce more results in developing gas lasers than in developing solid-state lasers.

Because of the problems in making and characterizing solid-state laser materials, developers have concentrated on a few types with properties that meet the needs of funding agencies. The neodymium laser has benefited from years of optimization, while other promising types have received much less attention. In recent years, there has been growing interest in alternative solid-state lasers, and that has led to some new types discussed later in this chapter.

Heat Conduction and Optical Distortion

Another vital role of the host material is to conduct away waste heat left over from laser action. Like other lasers, solid-state lasers are inherently inefficient. Typically only about one percent of the excitation energy emerges as a laser beam. Some of this energy is lost in the electrical power supply that drives the light source and in the light source itself, but some of it ends up in the solid-state laser material. This makes thermal conductivity of the laser host a very important concern.

Heat accumulation in solid-state laser materials can be bad news in three ways:

1. Excess heat can damage the laser material itself, causing it to warp, crack, or soften.

2. Temperature changes influence population distributions and gain characteristics of the light-emitting species, usually decreasing gain, laser efficiency, and output power.

3. Thermal expansion of the laser material can create refractive-index differentials in the laser rod, which can bend light within the rod so the light does not oscillate properly between the laser cavity mirrors. This increases losses and decreases output power.

The first of these problems is the most dramatic, but when it occurs it may be a consequence of the second and third, which decrease laser efficiency and hence increase waste heat generated.

In practice, the need to avoid these problems limits the number of materials that can serve as solid-state laser hosts. Certain crystals offer the best heat conductivity, but some glasses also are used because they can be made into large rods and blocks, as we will see later. Even using these materials, solid-state lasers usually require some type of active cooling, with forced air or flowing water, except when they are operated at modest output powers.

Laser Geometry

The archtypical solid-state laser is a small rod about the shape of a round pencil, but usually a little shorter. The small rod diameter helps ease heat-dissipation problems and, as we will see later, it can generate an impressive amount of light. Mirrors typically are mounted at either end of the laser rod, as shown in *Figure 7-1.*

There are limits to the amount of power available from a single-rod laser oscillator, so some high-power lasers use one or more external amplifier stages, as shown in *Figure 7-3*. While the first stage of such an oscillator-amplifier has mirrors on both ends (one of which is the partly transparent output mirror), the second and any subsequent stages do not. The beam makes a single pass through their volume, extracting stored light energy accumulated in a population inversion. Note that in some cases, such as in the slab amplifier, the beam may bounce around within the amplifier so it passes through most of its volume, but it does not oscillate back and forth over the same path. The mirrors bend the optical path so it stays on the page in the diagram; mirrors are added to real laser amplifiers so the beam stays on the optical table.

As you can see from *Figure 7-3*, solid-state lasers can take various shapes besides rods. The one at the upper right that looks

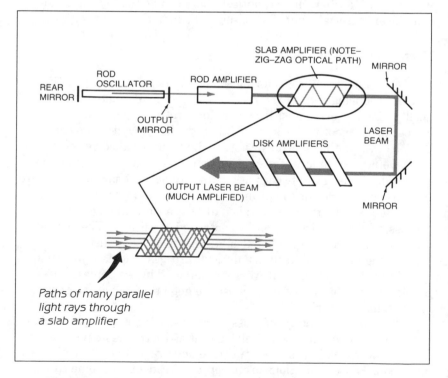

Figure 7-3. Laser oscillator and amplifiers.

like a parallelogram is called a "slab" amplifier, because of its shape. Refraction at the surface bends the beam so it bounces along its sides. If we expanded the input beam to cover the entire input of the slab laser, we would have light rays following many parallel paths through the slab and collecting light energy from most of its volume, as shown at the bottom of the illustration. The disk amplifiers are flat plates. They are used when a large-diameter beam is needed because heat can be dissipated through their thin sides more readily than it could be through a large solid chunk of material.

The larger chunks of laser material have the advantage of storing more energy, so they can generate higher-energy pulses. However, they also dissipate heat more slowly through their larger volume, so they must be operated at a much lower repetition rate than smaller rod lasers that produce lower-energy pulses.

Reduce the diameter of a solid-state laser rod enough and you wind up with a fiber. In fact, optical fibers can be made from solid-state laser materials, but fiber lasers remain in the laboratory.

Material Quality and Size

A final factor in considering solid-state laser materials is the ease of producing rods of the required size with adequate optical quality. The material must have a uniform refractive index, so light can oscillate smoothly back and forth in the laser cavity. The rod must be free of flaws that scatter or absorb light. (Light absorption at such flaws would lead to localized heating and could crack or shatter the laser rod.) Some crystals with otherwise attractive characteristics are too difficult to grow or too prone to flaws to be of any practical use.

The growth of laser crystals is at best a slow and agonizing process, which requires carefully controlled conditions. The crystals are grown in blocks called "boules," from which rods must be drilled. The cutting of the rod itself is a difficult process, and can sometimes reveal flaws in crystals that were not detected in the boules. These problems must be traded off against the good thermal characteristics of crystalline materials.

The main advantage of glass is the ease of producing large slabs of uniform optical quality. Offsetting that attraction is its poor heat dissipation, which limits repetition rate and prevents continuous-wave operation. Those limitations are acceptable tradeoffs in some cases, where the advantages of the higher pulse powers available from larger-size glass amplifiers offsets the lower repetition rates of glass lasers.

OPTICAL PUMPING AND SOURCES

Because solid-state laser materials are nonconductive, the only practical way to produce a population inversion is to illuminate the laser material with a bright light. Photons raise the light-emitting species to an excited state that leads to the laser transition. The choice of the source depends on the nature of the laser material.

Pumping Bands and Absorption

We mentioned earlier that all materials have characteristic energy levels and transitions. Each atomic and molecular species emits light on certain transitions when they drop from excited states. Likewise, they absorb light at characteristic wavelengths when they are in the ground state or other low levels. This absorption is the essential part of optical pumping.

Absorption can be at a narrow or a broad range of wavelengths, depending upon the transitions involved. Conceptually, a narrow range of wavelengths may seem simpler, because they involve a transition between a simple pair of energy levels. However, that usually is not a good arrangement for optical pumping, because lasers are the only light sources that concentrate their output at such a narrow band of wavelengths. The problem with lasers, as we learned earlier, is that most types are not very efficient at converting input energy into light. Thus, laser-pumped lasers tend to be inefficient, although as we will see later semiconductor laser pumping of solid-state lasers is an exception.

A more efficient approach generally is to match a broadband pump source, such as a flashlamp, with an absorption that spans a broad range of wavelengths. *Figure 7-4* shows the absorption spectrum of neodymium-YAG (Nd-YAG), the most common solid-state laser medium. This version has been smoothed out; actual measurements show many sharp spikes. Flashlamps and arc lamps used to pump neodymium lasers emit a broad range of light in the visible, near-infrared, and near-ultraviolet wavelengths, much of which Nd-YAG can absorb. Because of their high intensity, reasonable light-producing efficiency, and ready availability, such lamps have been standard sources for pumping neodymium lasers. Semiconductor lasers made of gallium aluminum arsenide emit near-infrared light in a range that includes some of the strongest neodymium absorption lines. As available power levels increase, GaAlAs lasers are becoming popular pump sources for neodymium lasers

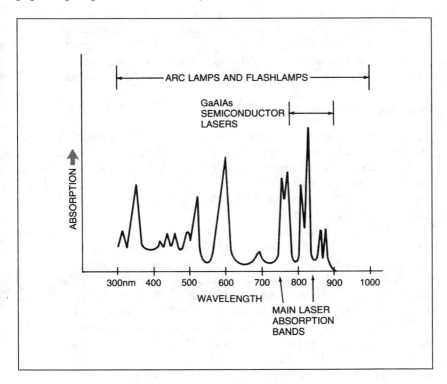

Figure 7-4. Broadband absorption of neodymium, compared with output of flashlamps, arc lamps, and semiconductor lasers.

Absorption spikes, which are hard to show in this version of the neodymium spectrum, come from transitions between well-defined energy levels. The broadband absorption comes from interaction of vibrational energy levels in the crystal with electronic energy levels of individual atoms. The vibrational transitions involve small enough energy changes that they can combine with electronic transitions to absorb light over a continuous range of wavelengths, such as near 800 nm in the near infrared. This broadband absorption makes it practical to pump solid-state lasers with conventional light sources such as flashlamps or arc lamps, which emit light over a wide range of wavelengths, and even makes it easier to match the wavelengths of the few lasers that can make efficient pump sources. Note that the absorption bands normally are not as wide as the range of wavelengths available from pump lamps, but that they do allow absorption of much of the pump light.

Flashlamp Pumping

The first laser was pumped with the bright flash of light from a xenon flashlamp, and flashlamp pumping remains common today. The extremely intense pulse of light from a flashlamp can excite most of the available atoms, and thus produce a population inversion even in a three-level laser material like ruby.

Flashlamps come in various shapes. Maiman used a helical lamp, shaped like a spring, but the most common type used to pump lasers today is a long linear tube. The lamp is filled with a gas such as xenon, and contains electrodes at each ends. Applying a brief high-voltage electrical pulse between the ends of the tube makes the gas break down electrically and conduct current, emitting a bright flash of light. The whole process requires about a thousandth of a second. Typically, the pulse from a solid-state laser will be slightly shorter than the flashlamp pulse, because the flashlamp does not instantaneously generate enough power to raise the laser above threshold.

Flashlamp pumping is common for pulsed solid-state lasers. It is limited in repetition rate by the switching electronics and the lamp itself. If desired, shorter pulses can be produced by Q switches, as described in Chapter 5.

Arc Lamp Pumping

Continuous-wave solid-state lasers can be pumped with electric arc lamps, in which a steady electric current flows through a gas-filled tube, producing intense light. This light is bright enough to sustain a continuous population inversion in some solid-state laser materials which are capable of continuous laser operation. (Not all solid-state laser materials can operate continuous wave.) In practice, lasers pumped with arc lamps can be pulsed with Q switches, modelockers, or cavity dumpers. This allows pulsing at higher repetition rates than is possible with flashlamps.

Laser Pumping

In general, laser pumping of lasers is not efficient, and is used as a last resort, when it is the only way to excite a certain type of laser. However, semiconductor lasers are so compact and efficient that they are an exception to the rule, and make better pump sources for some low-power solid-state lasers than flashlamps.

Semiconductor lasers are inherently among the most efficient types. So far, the only type of semiconductor laser used for optical pumping is gallium-aluminum arsenide (GaAlAs), which emits at 750 to 900 nm in the near-infrared. Those wavelengths fall within the absorption bands of some important solid-state laser materials, particularly neodymium, as you can see in *Figure 7-4*. Because the semiconductor laser wavelength falls at a strong neodymium-YAG absorption band, little of it is wasted. This is not the case with flashlamps and arc lamps; as you can see in *Figure 7-4*, much of their light falls at wavelengths where neodymium-YAG has little or no absorption.

For many years, semiconductor laser pumping of solid-state lasers was not taken seriously because semiconductor lasers did not produce high enough output powers. In recent years, however, developers have made semiconductor laser arrays that can emit steady powers of about a watt, as described in Chapter 8. Groups of such laser arrays can deliver enough pump power to excite a solid-state laser. Their high overall efficiency and the small amount of waste heat permits design of neodymium-YAG lasers that are smaller and more efficient than flashlamp-pumped versions. So far output powers from individual arrays remain modest, but average power levels have reached the one-watt level and are increasing. Higher powers can be obtained simply by ganging together several arrays to pump a single Nd-YAG rod.

Pumping Geometries

Laser designers no longer use flashlamps of the helical or coiled-spring design like those used in early solid-state lasers. The most common types are linear lamps. *Figure 7-5* shows three common ways to transfer light from a linear flashlamp (or arc lamp) to a laser rod, along with a sketch showing a laser rod slipped inside a helical lamp.

Figure 7-5B shows how linear lamps are placed beside each other in a reflective cylinder. An end view of this arrangement is given in *Figure 7-5C*. Two other ways of coupling light from linear lamps are shown only in end view. One is to put the lamp and rod at the two foci of a reflective elliptical cavity (*Figure 7-5D*). Light radiating from one focus of the cavity is reflected to the other focus from any point on the ellipse. If the pump lamp is at one focus, this means that the reflective elliptical cavity focuses all its light onto the laser rod at the other focus.

The same principle can be used in a dual-elliptical cavity. The laser rod is put at the common focus of two overlapping ellipses. A lamp is at the other focus of each ellipse, so the light from both lamps is focused onto the laser rod, as shown in *Figure 7-5E*. (Although the figure implies the cavity is filled with air, in fact some pump cavities are filled with flowing water that removes waste heat from the laser.)

Different geometries are used for laser pumping. Typically, semiconductor lasers are arranged along the side of the laser rod, so they illuminate its entire length. However, if pumping is with a single external laser, the pump beam typically is directed along the length of the laser rod. This approach has some drawbacks because of the need for cavity optics that behave differently at the pump wavelength than at the emission wavelength, and because the amount of pump light drops with distance through the laser rod. However, it does concentrate the light from a single pump laser into the laser rod.

RUBY LASERS

Ruby was the first laser material, but it is far from an ideal one. The three-level ruby laser system is inherently less efficient than the four-level neodymium laser, and is limited to pulsed operation at low repetition rates. However, ruby remains important for a limited number of applications that require its high-power red pulses.

Laser Medium and Physics

Natural ruby is a gemstone, but ruby lasers use a synthetic ruby made by doping aluminum oxide (which crystallizes to form sapphire) with 0.01 to 0.5% chromium. Sapphire is naturally clear, but the chromium atoms give it a pink or reddish color.

7

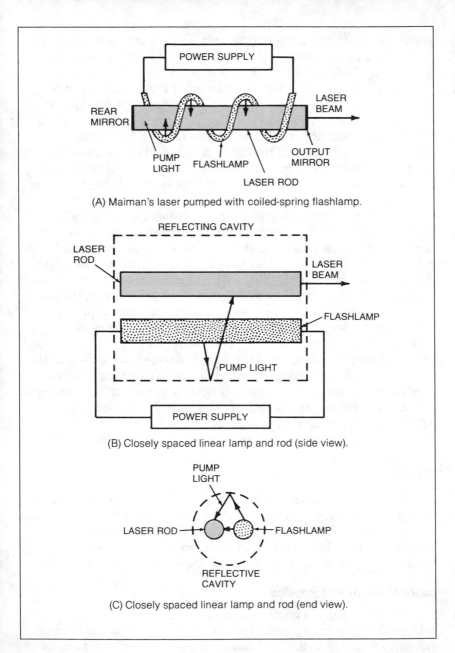

(A) Maiman's laser pumped with coiled-spring flashlamp.

(B) Closely spaced linear lamp and rod (side view).

(C) Closely spaced linear lamp and rod (end view).

Figure 7-5. Flashlamp

Ruby is a three-level laser system, with its energy-level structure shown in *Figure 7-6*. Ground-state chromium atoms absorb light in two pump bands, one centered near 550 nanometers, the other at about 400 nm. These are well-matched to the output of xenon flashlamps. After about 100 nanoseconds, the excited chromium atoms release energy as vibrations of the crystalline lattice and drop to a pair of closely spaced metastable levels. (They are too closely spaced to show as separate in the illustration.) These metastable states have lifetimes of 3 milliseconds at room temperature, long enough to serve as the upper laser level of a 694.3-nm laser transition to the ground state.

The three-level energy structure makes ruby relatively inefficient, with typical wall-plug efficiency 0.1 to 1%. Because the laser level terminates with the ground state, ruby that has not been excited can absorb the 694.3-nm laser line. This makes it essential to pump all of a ruby rod to avoid excess absorption. However, a ruby rod can store

(D) Simple reflective elliptical cavity (end view).

(E) Dual elliptical cavity (end view).

pumping geometries.

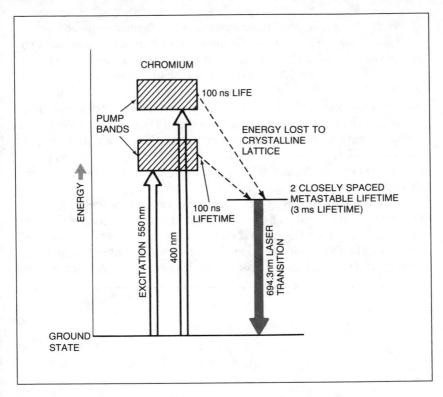

Figure 7-6. Energy levels of chromium atoms in ruby laser.

more energy than neodymium, and can readily be Q-switched to produce short pulses with energies of a few joules. It also can be used in an oscillator-amplifier configuration to raise total laser energy.

Ruby conducts heat well, and resists damage from excess optical energy as long as its surface is kept clean of dirt that could absorb excess light and heat the material. However, because it is a three-level laser, its laser properties degrade rapidly with increasing temperature, so it must be operated at low repetition rates (typically no more than a few pulses per second except for very small rods) to prevent heat build-up.

Practical Ruby Lasers

Typically ruby rods are 3 to 25 mm (⅛ inch to 1 inch) in diameter, and up to about 20 cm (8 inches) long. Without Q switching, energy in a multiple-transverse mode pulse can reach 100 joules in a millisecond, although typically it is much lower. Pulse energies are much lower if oscillation is limited to TEM_{00} mode. Q-switching can compress pulse duration to 10 to 35 nanoseconds. That limits pulse energy to a few joules, but gives peak power in the 100-megawatt range.

Ruby lasers often operate in a dual-pulse mode for holographic measurements. In this mode, a Q switch generates two short pulses during a single long flashlamp pulse. In other words, the Q switch turns the laser on twice during the millisecond duration of the flashlamp pulse. The two pulses record holograms on the same film, allowing comparison of the effects of stress on an object.

Ruby lasers require active cooling with forced air or flowing water, but even then their repetition rate is limited to low levels except for the smallest rods.

NEODYMIUM LASERS

The most important solid-state lasers now are types in which the light-emitting species is the rare earth neodymium in a glass or crystalline matrix. The two types of neodymium lasers share a common energy-level structure, but they differ in some ways because of the properties of the different hosts. The most common host is yttrium-aluminum garnet, a hard, brittle crystal generally known by the acronym "YAG". Glass is the second most useful host for neodymium. Dozens of other neodymium hosts have been tested, but only a handful have been developed extensively. The two most important alternatives are yttrium lithium fluoride (YLF) and yttrium aluminate (YALO, from its chemical formula $YAlO_3$). The main wavelength of neodymium lasers is 1064 nanometers in Nd-YAG, and at slightly different wavelengths in other hosts. However, for most practical purposes there is no real difference among these wavelengths.

The main energy levels in the four-level neodymium laser system are shown in *Figure 7-7*. In this general diagram, we round the wavelength to 1.06 micrometers (equivalent to 1060 nm) to indicate that we are talking about the entire family of neodymium lasers, not a specific type. Interactions between neodymium atoms and the host material can change wavelength by about one percent. The two primary pump bands shown are in the 700–850 nm range, which raise neodymium atoms to one of two broad high-energy bands. However, as indicated in *Figure 7-4*, there also are other important absorption lines. Traditionally, neodymium lasers have been pumped by flashlamps or arc lamps. However, GaAlAs semiconductor lasers—which emit near 800 nm—have become increasingly attractive pump sources as their output powers have increased.

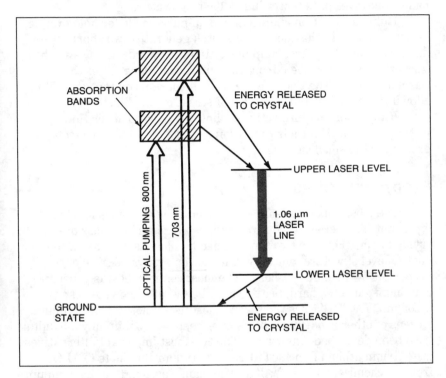

Figure 7-7. Major energy levels in 1.06-micrometer neodymium laser.

The optically excited neodymium atoms quickly decay to the metastable upper laser level, releasing their excess energy to the crystalline lattice. In this state, they are stimulated to emit on the main 1.06-micrometer laser transition, dropping to a lower laser level which they quickly leave, again by transferring energy to the crystal.

Our energy-level diagram in *Figure 7-7* simplifies the situation by omitting the splitting of the upper and lower laser levels in neodymium. This splitting creates several other laser transitions in the near infrared, but all of them are weaker than the main 1.06-μm line. The strongest, at 1318 nm in YAG, can produce about 20% as much power as the 1064-nm YAG line.

Neodymium-YAG

In neodymium-YAG, neodymium is an impurity that take the place of some yttrium atoms in the YAG crystal. YAG's chemical formula is $Y_3Al_5O_{12}$; its crystalline structure is similar to that of garnet. The crystal has good thermal, optical, and mechanical properties, but it is hard to grow. The crystal is grown in blocks called "boules" from which rods are drilled. Typical YAG rods are 6 to 9 mm (0.24–0.35 inch) in diameter and up to 10 cm (4 inches) long.

The small size of Nd-YAG rods and the optical properties of neodymium atoms limit the amount of energy that can be stored in a typical rod to about half a joule, far less than in a ruby rod. However, most of that energy can be removed from the rod in a Q-switched pulse, and the energy can be replenished quickly—in well under the millisecond duration of a flashlamp pulse—so a repetitively pulsed Nd-YAG rod can generate high average powers, as well as high peak power in Q-switched pulses.

The thermal and optical properties of Nd-YAG let it be pumped either continuously with an arc lamp or by a series of flashlamp pulses. Maximum average power from an Nd-YAG laser can reach hundreds of watts, although most operate at much lower powers. The peak power can reach tens or hundreds of kilowatts in a millisecond-long pulse generated by a flashlamp, or over 100 MW in a Q-switched pulse of 10 to 20 nanoseconds.

When doped in YAG, neodymium has a high gain, and can be operated as an oscillator with a stable or unstable resonator. Although energy storage in a rod is limited, neodymium lasers can be operated in an oscillator-amplifier configuration to reach higher powers. The oscillator and amplifier need not use the same host material—often, oscillators are Nd-YAG or another crystal, and the oscillators are Nd-glass.

Neodymium-Glass Lasers

The principal attraction of glass as a solid-state laser host is the well-developed technology for making large chunks of laser glass with good optical quality. Also, because neodymium-glass has lower gain than Nd-YAG, a glass laser rod can store more energy than an equal-sized YAG rod. Thus, Nd-glass lasers can generate higher-energy pulses than YAG lasers. The principal tradeoff is that glass has poorer thermal characteristics, so more time is needed for it to cool between pulses. Thus, glass laser oscillators cannot be pumped continuously, and normally operate at much lower repetition rates than Nd-YAG lasers.

Laser glass can be made in a much wider variety of shapes than Nd-YAG. For example, the Nova laser used in fusion research at the Lawrence Livermore National Laboratory includes both glass rod amplifiers up to 5 centimeters in diameter and disk amplifiers up to 46 by 85 cm (18 by 33.5 inches) across. As shown in *Figure 7-3*, the disk amplifiers are thin, to assist in heat removal, and tilted at an angle, to avoid potentially harmful reflections. Laser glass also can be cast in slabs, which can amplify the light from laser oscillators, and also are shown in *Figure 7-3*.

Neodymium-glass lasers emit broader bandwidth light than Nd-YAG. This is significant in modelocking, because it lets you generate shorter pulses with Nd-glass than with Nd-YAG. The wavelength emitted depends on the glass composition; it is 1062 nm for silicate glass, 1054 nm for phosphate glass, and 1080 nm for fused silica.

Neodymium Laser Configurations

Neodymium lasers are pumped in the same way as other solid-state lasers, using flashlamps, arc lamps, or semiconductor lasers. The pumping arrangements used with flashlamps and arc lamps were shown in *Figure 7-5*.

The gain in neodymium lasers is high enough that they can use either stable or unstable resonators. An unstable resonator has the advantage of extracting laser energy from more of the rod. Close to the laser, some unstable resonators produce beams with a bright ring around a central point of minimum intensity, which looks like a ring or doughnut in cross section. However, far from the laser the hole vanishes to produce a bright central spot. A stable resonator can produce the standard gaussian TEM_{00} beam with a bright central spot, but it does not extract laser energy from as large a fraction of the rod volume. Unstable resonators have been growing in popularity because for most applications output power and energy are more important than near-field beam quality.

As we mentioned earlier, an external amplifier can boost the output power and energy from a neodymium oscillator. Oscillator and amplifier stages need not be made of materials based on the same host, but the host wavelengths must be close enough that the oscillator wavelength falls within the gain bandwidth of the laser amplifier.

Practical Neodymium Lasers

The neodymium laser is extremely versatile, and can take many forms. Some neodymium lasers are made for general laboratory use, while others are made for specific jobs including measuring the ranges to military targets, treating eye disease, or drilling holes. This wide range of uses means that neodymium lasers can look quite different from each other. The compact, battery-powered laser rangefinder or target designator designed for battlefield use is very different from a massive water-cooled high-power drilling laser built to stand on a factory floor. Yet flashlamp-pumped Nd-YAG rods lie at the core of both.

The wide range of applications leads to a range of design choices. General-purpose laboratory lasers are built to give the user as many options as possible. Typically, room is left in the laser cavity or on the outside of the laser to add accessories to change wavelength and/or pulse length. On the other hand, lasers designed for specific applications, such as drilling holes, may allow for few modifications not essential to performing their major job.

The overall efficiency of lamp-pumped neodymium lasers is in the same 0.1 to 1% range as ruby lasers. The smallest models can operate without active cooling, but larger types require either forced-air cooling (i.e., a fan), or flowing-water cooling.

Efficiencies can be much higher for semiconductor-laser pumping, leading to lower levels of waste heat. Combined with the inherent small size of semiconductor lasers, this makes it possible to build hand-held neodymium lasers pumped by semiconductor lasers. Because output of the semiconductor lasers is absorbed efficiently by the neodymium laser, designers need to pay more attention to removing waste heat from the semiconductor lasers than from the Nd-YAG rod.

Changing Wavelength and Pulse Length

We saw in Chapter 5 that a variety of accessories are available to change a laser's wavelength and pulse duration. Some of the most common uses of these accessories are with neodymium lasers.

Harmonic Generation

The near-infrared wavelength of Nd-YAG lasers is fine for some purposes, but visible or ultraviolet light is better for many others. Fortunately, neodymium lasers generate high enough powers (particularly in Q-switched pulses) that nonlinear harmonic generation can readily produce visible or near-ultraviolet wavelengths.

In Chapter 5, we showed how nonlinear interactions between light waves and certain materials could generate light at different wavelengths. The simplest such interaction is frequency doubling. Twice the frequency corresponds to half the wavelength, so frequency-doubling the 1064-nm output of a neodymium laser (*Figure 7-8A*) produces green light at a wavelength of 532 nm. Using slightly more complex optical arrangements, such as shown in *Figures 7-8B and C*, you can generate the third harmonic at 355 nm and the fourth harmonic at 266 nm, both in the ultraviolet. These wavelengths also have many applications. Note that some of the input wavelengths always are present, and must be removed by blocking with a filter, or separating them with a wavelength-selective beamsplitter.

Harmonic generation from neodymium lasers has become so common that some lasers are packaged with the harmonic generator inside. You hear people talking about "green" neodymium lasers, even though the neodymium atom does not emit green light. You should always remember that the neodymium laser's primary wavelength is the infrared region.

Changing Pulse Length

Without external pulse control, a flashlamp-pumped Nd-YAG or Nd-glass laser will produce pulses about a millisecond long. Those are usable for some purposes, including some welding and drilling. However, many other applications require pulses that are much shorter or have much higher peak power. Those can be generated by the Q-switching, cavity-dumping, or modelocking techniques described in Chapter 5.

With a flashlamp-pumped Nd-YAG laser, a Q switch can generate pulses lasting 3 to 30 nanoseconds, with repetition rate depending on that of the flashlamp. A Q switch also can generate short pulses from a continuously pumped Nd-YAG laser, but these pulses will be longer, often hundreds of nanoseconds, because of

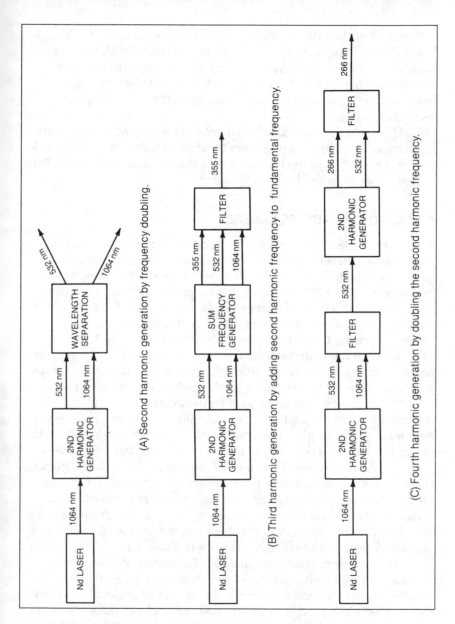

Figure 7-8. Generation of second, third, and fourth harmonics of an Nd-YAG laser.

differences in the rate of energy transfer and build-up within the laser. Cavity-dumped pulses also are shorter with flashlamp-pumped lasers than with continuous-wave models. However, the duration of modelocked pulses is the same for both (30 to 200 picoseconds) because it depends on the range of wavelengths emitted by the neodymium atoms.

Q switches in particular often are built into neodymium lasers that are sold as Nd-YAG lasers emitting short pulses with high peak power. Again, it pays to remember that the generation of these short pulses depends on the internal Q switch.

VIBRONIC SOLID-STATE LASERS

Vibronic lasers are closely related to other solid-state lasers, but have subtly different energy-level structures. The upper and lower laser levels in ruby and neodymium lasers are single states, as shown in *Figures 7-6* and *7-7*. However, in vibronic lasers the lower laser level actually is a band of energy levels, as shown in *Figure 7-9*. This band is a single electronic energy level that has a continuum of sublevels arising from vibrations of the crystalline lattice.

When an atom drops from the upper laser level (which may be a single electronic state or—as shown in *Figure 7-9*—the bottom of a band) to the lower band of laser levels, it changes both electronic and vibrational energy. This compound transition is called "vibronic," a contraction of vibrational-electronic. A vibronic transition can occur over a range of energies, because the excited atom can drop from the upper level to anywhere within the lower vibronic band. Normally laser emission is to one of the upper levels in the lower band, because they are less populated than those close to the ground state. This means that vibronic lasers can emit light at a comparatively broad range of wavelengths, which can vary by up to ± 20% from a central wavelength. The emission ranges for some important vibronic lasers are listed along with their compositions in *Table 7-1*.

The tuning ranges listed in *Table 7-1* represent the spectral ranges over which vibronic lasers can show laser gain—the gain bandwidths we discussed earlier. Without special optics in the laser cavity, vibronic lasers (like other types) emit light at the wavelength where they have highest gain, typically toward the middle of the range. However, if the cavity optics are adjusted so only certain wavelengths can oscillate in the laser cavity, as described in Chapter

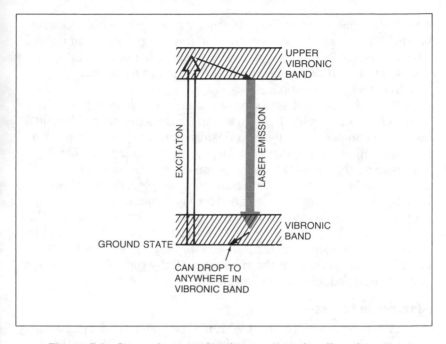

Figure 7-9. General energy level structure of a vibronic laser.

Table 7-1. Important Vibronic Lasers

Name	Composition	Output Range (nm)*
Alexandrite	Chromium-doped $BeAl_2O_4$	700–830
Co-MgF_2	Cobalt-doped MgF_2	1500–2300
Cr-GSGG	Chromium-doped $Gd_3Sc_2Ga_3O_{12}$	740–850
Emerald	Chromium-doped $Be_3Al_2(SiO_3)_6$	720–810
Ti-sapphire	Titanium-doped Al_2O_3	660–1060

*Approximate ranges; the exact tuning range depends on operating conditions and laser design

4, a vibronic laser will oscillate at other wavelengths within its gain bandwidth. The combination of tunable-wavelength output and the ability to generate wavelengths not available from other solid-state lasers has pushed development of vibronic lasers for military, research, and civilian applications.

Tunability is what sets vibronic lasers apart from other solid-state types. Otherwise, they work in much the same way. Vibronic lasers are pumped optically, by a flashlamp, an arc lamp, or another laser (including semiconductors). Depending on their pump bands, they may be able to use the same pump sources as ruby or neodymium lasers. Alexandrite rods, for example, can lase when inserted into laser cavities designed for neodymium rods, although they work better in cavities designed especially for alexandrite. This similarity should not be surprising, especially for alexandrite and the other chromium-doped materials, because they use energy levels closely related to those in the ruby laser. (Ruby rods also can work in cavities designed for neodymium rods.)

Alexandrite Lasers

Alexandrite was the first vibronic laser developed commercially, and it remains one of the more important types. The light-emitting species is chromium, added to $BeAl_2O_4$ (a mineral known as alexandrite) in concentrations of about 0.01 to 0.4%. Its energy levels are similar to those of the ruby laser, as shown in *Figure 7-10*, except for the existence of a vibronic band of the ground state. Alexandrite can emit on a 680.4-nm fixed-wavelength transition to the ground state that is equivalent to the 694.3-nm ruby transition, but it is not as efficient as ruby. Like ruby, alexandrite has pump bands at 380 to 630 nm which allow optical pumping with a flashlamp or arc lamp. Its strongest laser emission is at 700 to 830 nm.

Alexandrite has some peculiar kinetics that come from the fact that two electronically excited states together function as the upper laser level. One is the bottom of a vibronic band of energy levels, the other is a fixed state with longer lifetime and only slightly less energy. This combination makes the gain of alexandrite increase with temperature although the gain of most other lasers drops as temperature rises.

The lower laser level is a band of energy levels that are vibrationally excited states of the ground electronic level. When the chromium atom emits a laser photon, it drops to one of the vibrational

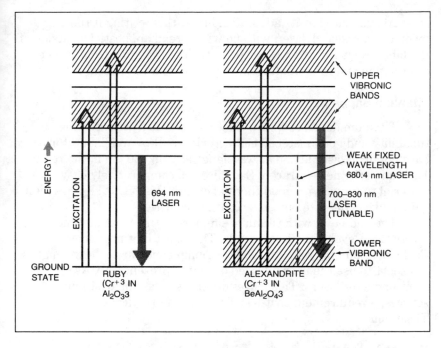

Figure 7-10. Comparison of ruby and alexandrite energy levels.

states, then releases the extra energy (above the ground state) as a vibration in the crystal lattice. As the laser temperature increases, the steady-state population of the vibrational sublevels slightly above the ground state increases. This makes it increasingly difficult to produce a population inversion with respect to these low-lying levels, which correspond to the higher-energy end of alexandrite's tuning range. Thus, the shortest wavelength available from alexandrite (and other vibronic lasers) tends to increase with temperature.

Alexandrite has lower gain than neodymium lasers, so more care must be taken in cavity design. On the other hand, this lets an alexandrite rod store more energy than an equal-sized Nd-YAG rod. (Alexandrite rods are similar in size to Nd-YAG rods.) Like Nd-YAG, alexandrite can operate pulsed or continuous wave. In pulsed operation, average powers can reach 100 watts, lower than the most powerful Nd-YAG lasers, but enough to make it among the more powerful lasers available.

Alexandrite has its advocates, but at this writing it remains primarily a research laser, with no widespread applications outside of the laboratory. However, it has potential uses ranging from target designation and rangefinding on the battlefield to medical treatment.

Titanium-Sapphire

Titanium-doped sapphire has two major attractions: wide tunability and good material characteristics. Research laboratories have obtained laser action at wavelengths from 660 nm in the red to 1160 nm in the near-infrared. Sapphire is a comparatively easy material to grow, with good thermal characteristics. It also has high gain, and can operate pulsed or continuous wave.

The bad news with titanium-sapphire is the short lifetime of its upper laser level, 3.2 microseconds. This short lifetime makes flashlamp pumping difficult, and the pump bands are at wavelengths too short to use the near-infrared output of present GaAlAs semiconductor lasers. This leaves other lasers as the best pump sources, a requirement that has discouraged practical use of Ti-sapphire.

Garnet Materials

Crystals in the garnet family (other than YAG) have good thermal and optical properties that make them attractive hosts for solid-state vibronic lasers. So far the best-investigated type is gadolinium scandium gallium garnet, known as GSGG, a crystal with chemical formula $Gd_3Sc_2Ga_3O_{12}$. This material has been doped with chromium as a vibronic laser, as well as with neodymium. Laser action also has been demonstrated in double-doped crystals, Cr-Nd-GSGG. However, important practical problems remain, including the difficulty in pumping with a flashlamp and concerns about the high cost of scandium.

OTHER SOLID-STATE LASERS

Many other solid-state lasers have been demonstrated in the laboratory, but only a handful are practical enough to be available commercially. There are three principal types: holmium, erbium, and color-center lasers. The first two are analogous to neodymium lasers; the third is a type by itself.

Erbium Lasers

Erbium is a rare earth like neodymium. It has several laser lines in the infrared. The most important are:

- 0.85 μm
- 1.23 μm
- 1.54 μm
- 1.73 μm
- 2.9 μm

Erbium can operate in hosts including yttrium-lithium fluoride (YLF), YAG, and glass. Like neodymium, it produces fixed wavelengths and can be pumped with a flashlamp. However, it is not as well developed as the neodymium laser, and so far its applications have been limited to those which require specific wavelengths. One is the 1.54-μm wavelength, sought for military training equipment because it poses little hazard to the human eye (see Appendix A for a primer on laser safety). Lately, developers of surgical lasers have become interested in the 2.9-μm wavelength because it is strongly absorbed by tissue and unlike other lasers can cut bone.

Holmium Lasers

Holmium is another rare-earth element which can lase when doped in YLF or YAG. Its best output wavelength is 2.06 micrometers, but it has found few applications because demand for that wavelength is limited. Another drawback has been that it works best when cooled to the 77 K temperature of liquid nitrogen. The thermal energy at room temperature is enough to populate the laser's lower level, making it a three-level laser, but at cryogenic temperatures the population in the lower level drops, and holmium becomes a four-level laser. Applications remain limited.

Color-Center Lasers

Color-center lasers work in a quite different way than other solid-state lasers. Their active media are crystals that are doped with impurities that introduce flaws in the crystalline lattice. The flaws themselves absorb and emit light.

The simplest way to envision the working of a color-center laser is as a group of atoms that fit in a hole in a crystalline lattice, as shown in *Figure 7-11*. At least one of the atoms in the hole is an impurity of different size than the atoms that make up most of the crystalline lattice. The atoms can fit into the hole in different

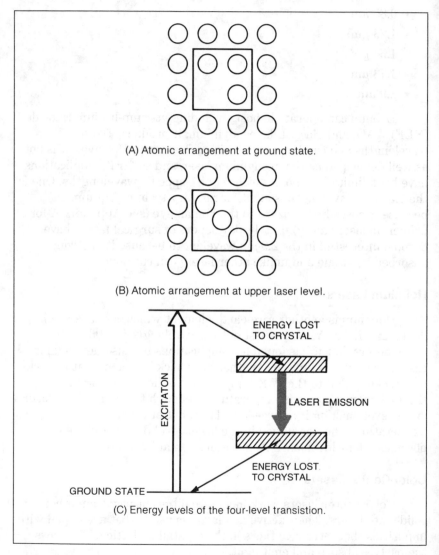

(A) Atomic arrangement at ground state.

(B) Atomic arrangement at upper laser level.

(C) Energy levels of the four-level transistion.

Figure 7-11. The color-center laser.

arrangements, each with a characteristic energy. When light shines on the crystal, it can raise the atoms from their ground state to a higher-energy state. They then slip to a metastable state with less energy, but still well above the ground state. This metastable state serves as the upper laser level, which can be stimulated to emit laser light when the group of atoms drops to a lower energy level. (By the way, our example only shows the general idea, not a specific color-center-laser configuration.)

The energy levels in a color-center laser are not as sharp as in isolated atoms. Some of the energy from changes in the atomic configuration goes into strain or vibration in the crystal. This lets color center lasers both absorb and emit light over a broad range of wavelengths. In practice, color-center lasers must be pumped with laser light, but they have gain over a broad range, and thus can be tuned in wavelength. Commercial color-center lasers emit near-infrared light, either near 1.5 micrometers, or between about 2.3 and 3.3 micrometers.

The need for a costly pump laser and for cryogenic cooling to produce the flaws called "color centers" limit the practical applications of color-center lasers. However, their ability to produce a narrow-line, tunable output not otherwise available at those wavelengths had led to some applications.

7

What We Have Learned

- Light-emitting atoms in solid-state lasers are fixed in a crystal or glassy material, and excited by light from an external source.

- Semiconductor lasers are not considered "solid-state" types.

- The element that emits light in a solid-state laser accounts for only a small fraction of the crystal or glass substance of the laser rod. The most common light emitters are the ions Cr^{+3}, Nd^{+3}, Er^{+3}, and Ho^{+3}.

- Transparency, thermal conductivity, and interaction with the light emitting species are important features of host materials.

- It is harder to develop new crystals than new gas lasers, so solid-state laser development has concentrated on fewer materials.

- Excess heat can damage laser media, decrease laser gain, or cause optical distortion in the rod. Optical distortion can impair laser action.

- Solid-state laser rods usually are about the size of a round pencil.

- Solid-state laser amplifiers can be rods, slabs, or disks.

- Crystal growth is a major practical concern in solid-state lasers. Glass is the easiest material to produce in large sizes.

- Flashlamps of various shapes remain common pumping sources for solid-state lasers because the intense flash can readily produce a population inversion. Linear pump lamps are used in reflective cavities that focus light onto the laser rod.

- Arc lamps pump some continuous-wave solid-state lasers.

- Semiconductor lasers can be efficient pumps for some solid-state lasers, but other laser pumping is inefficient.

- Ruby is a three-level laser system, but remains in use despite its inherent inefficiency because it can produce high-power red pulses at 694 nm.

- Energy in a ruby-laser pulse can reach 100 joules. Q-switched pulses of 10 to 35 nm can have peak power to 100 MW.

- The most important solid-state lasers are 1.06-micrometer

neodymium types, with the host either a crystal known as YAG or glass.

- Nd-YAG has good thermal, optical and mechanical properties, but it is hard to grow. Nd-YAG can operate continuous wave or pulsed.

- The main attraction of glass as an Nd host is the ability to produce large blocks with good optical quality.

- Semiconductor laser pumping of neodymium lasers is more efficient than flashlamp pumping.

- Harmonic generation can convert 1064-nm Nd-YAG output to 532, 355, or 266 nm.

- Nd-YAG lasers often are Q-switched to generate nanosecond pulses with high peak power.

- The lower laser levels of vibronic lasers are bands broad enough that such lasers can be tuned over a range of wavelengths. Tunability sets vibronic lasers apart from other solid-state lasers.

- Alexandrite, like ruby, emits on lines of chromium. Its output wavelengths are 700–830 nm. Its tunability comes from the vibrational band that is its lower laser level.

- Titanium-sapphire can be tuned from 660–1160 nm and is easy to grow.

- Garnets other than YAG are promising hosts for vibronic lasers.

- Erbium-doped crystals and glasses have several near-infrared laser lines.

- Holmium's best output wavelength is at 2.06 μm.

- Color centers are flaws in crystals that can emit laser light tunable in the near-infrared.

WHAT'S NEXT?

In Chapter 8, we will learn about one of the fastest-moving areas of laser technology, semiconductor lasers.

7

Quiz for Chapter 7

1. A host material for a solid-state laser must meet which of the following criteria?
   a. Must be transparent at the pump wavelength
   b. Must be transparent at the laser wavelength
   c. Must be able to conduct away waste heat
   d. A & B only
   e. A, B, and C

2. Which of the following is the best pump source for a solid-state laser?
   a. A flashlight
   b. A flashlamp
   c. A helium-neon laser
   d. An electrical discharge
   e. A fluorescent tube

3. What type of laser is the most efficient pump for neodymium lasers?
   a. GaAlAs semiconductor
   b. Argon-ion
   c. Helium-neon at 632.8 nm
   d. Ruby
   e. InGaAsP semiconductor

4. Which of the following optical pumping arrangements is no longer used with solid-state lasers?
   a. Semiconductor-laser pumping from the sides of the rod
   b. Pumping with a close-coupled linear flashlamp
   c. Pumping with a linear flashlamp in an elliptical cavity
   d. Pumping with a helical (spring-like) flashlamp surrounding the rod
   e. All of the above are widely used

5. What are the pump bands of ruby lasers?
   a. 750–900 nm
   b. One at 400 nm, another at 550 nm
   c. 694.3 nm
   d. 1064 nm
   e. None of the above

6. What makes it feasible to pump a neodymium laser with semiconductor lasers?
   a. High efficiency of semiconductor lasers
   b. Increased power available from semiconductor lasers
   c. Pump bands coincide with semiconductor laser output
   d. A and B
   e. A, B, and C

7. In which of the following characteristics is neodymium-doped glass better than Nd-YAG?
   a. Thermal characteristics
   b. Higher repetition rate
   c. Ease of producing large blocks
   d. Higher laser gain
   e. Much shorter output wavelength

8. Which of the following wavelengths cannot be readily generated from an Nd-YAG laser?
   a. 266 nm
   b. 355 nm
   c. 477 nm
   d. 532 nm
   e. 1064 nm

9. What differentiates vibronic lasers from other solid-state lasers?
   a. Much broader gain bandwidths and tunable output
   b. Higher gain
   c. Shorter wavelengths
   d. Can be pumped efficiently with semiconductor lasers
   e. All of the above

10. Which of the following lists includes only vibronic lasers?
   a. Nd-YAG, Ho-YLF, alexandrite, ruby
   b. Ruby, Alexandrite
   c. Titanium-sapphire, ruby, color-center lasers
   d. Titanium-sapphire, alexandrite, Cr-GSGG
   e. Nd-glass, Erbium-glass, gallium-aluminum-arsenide

Semiconductor Lasers

ABOUT THIS CHAPTER

Semiconductor laser technology has been growing explosively for the past several years. It is based on a combination of optical and semiconductor electronic technology that gives semiconductor lasers their own distinct properties. At first able to provide only modest powers, semiconductor lasers now are packaged in arrays that can deliver steady powers of a watt or more. In this chapter we will explore the principles of semiconductor lasers, and learn about the diverse types now available.

EVOLUTION AND BASIC CONCEPTS

The roots of semiconductor laser technology go back to the 1950s, when semiconductor physics was new. As far back as 1953, noted physicist John von Neumann considered the possibility of light amplification by stimulated emission in semiconductors, but he never formally proposed the idea. In 1957, Yasushi Watanabe and Jun-ichi Nishizawa applied for a Japanese patent on a "semiconductor maser" concept. The most detailed proposals for semiconductor lasers emerged in 1961 from Nikolai Basov's group at the Lebedev Physics Institute in Moscow. Basov and another Russian maser-laser pioneer, Aleksander Prokhorov, later shared the 1964 Nobel Prize in Physics with Charles Townes.

All these early ideas were based on phenomena that occur when a current flows through a junction between two segments of semiconductor material with different doping. If this current flows in the right direction through certain materials, it raises electrons into excited states that can eventually emit light. We'll explain that later in more detail.

In 1962, four independent groups in the United States succeeded in making semiconductor lasers within weeks of each other. (The winner, in a photo finish, was Robert N. Hall of General Electric Research and Development Laboratories in Schenectady, NY.) Those lasers stimulated tremendous interest, but they only worked when high-current pulses passed through them, and they required cooling to the 77 kelvins temperature of liquid nitrogen. It was not until 1970 that physicists at AT&T Bell Laboratories made the first semiconductor lasers able to produce a continuous-wave beam at room temperature. More years passed before semiconductor lasers were able to operate for thousands of hours at room temperature without self-destructing.

The state of the art in semiconductor lasers has continued to advance at an amazing rate. New arrays of many semiconductor laser stripes on a single chip can produce continuous powers of more than one watt. Millions of inexpensive semiconductor lasers are mass-produced for use in compact disc audio players. Sophisticated semiconductor lasers can transmit billions of bits per second through optical fibers.

Some tough technological problems do remain. One is continuous-wave output in most of the visible spectrum. Semiconductor lasers that emit wavelengths shorter than the red end of the visible spectrum tend to quickly self-destruct. Another is economical production of lasers that emit narrow ranges of wavelengths for fiber-optic communication systems. We will learn why as we explore the fundamentals of semiconductor lasers.

PROPERTIES OF SEMICONDUCTORS

A semiconductor is a material with electrical properties intermediate between those of a conductor and an insulator. The electrons in a conductor (e.g., a metal) are free to move in the material and carry electrical current. The electrons in an insulator are tightly bonded to atoms, and thus cannot move. In a semiconductor, electrons are loosely bonded to atoms, so some movement is possible.

It may be easiest to understand the nature of semiconductors by looking at the energy-level diagram in *Figure 8-1*. A pure semiconductor crystal such as silicon or germanium has exactly enough electrons to fill all the niches in the "valence" band. At room temperature, a few electrons are promoted to the "conduction" band, where they can freely move about. The number of electrons depends on the size of the gap and the temperature, according to the Boltzmann law:

$$N_2 / N_1 = \exp[-(E_2 - E_1) / kT]$$

where,

N_2 and N_1 are the numbers of electrons in the conduction and valence bands, respectively,

$E_2 - E_1$ is the energy difference between the two states,

k is the Boltzmann constant,

T is the temperature (in Kelvins).

This formula is the same one we used in Chapter 3 to describe populations of energy levels in atoms or molecules.

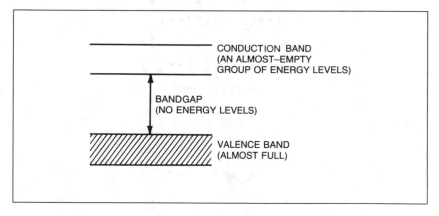

Figure 8-1. Valence band, conduction band, and bandgap in a pure semiconductor.

At room temperature there aren't many electrons in the valence band, but there are some. As we have seen from the Boltzmann equation, the number depends on the difference in energy between the valence and conduction bands—a quantity called the bandgap.

Doping Semiconductors

The simplest pure semiconductors are silicon and germanium, elements with four electrons in their outer shells. Each of those outer electrons goes to form a bond with another atom in the crystal, as shown in *Figure 8-2*.

The small number of electrons in the conduction band in a pure semiconductor can be increased by adding impurities which have different numbers of electrons in their outer shells. If the impurity is

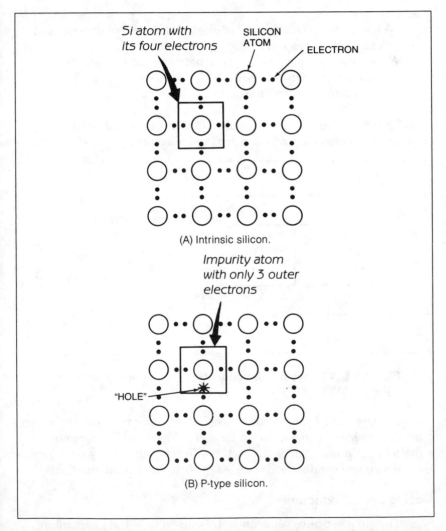

(A) Intrinsic silicon.

(B) P-type silicon.

Figure 8-2. Bonding in pure

present in only small concentrations, it fits into the crystal as if it was an atom of the semiconductor. In *Figure 8-2A*, the silicon atoms that form the bulk of the crystal each have four outer electrons that form bonds with adjacent atoms. If the impurity atom has five electrons in its outer shell, four of them form bonds with the surrounding silicon atoms, and the fifth is left free to conduct electricity in the crystal, as shown in *Figure 8-2C*. A semiconductor doped with such electron donors is called an *n-type* semiconductor, because the doping produces negative current carriers. Elements such as phosphorous or arsenic are used as electron donors in n-type silicon.

Something more complicated happens if the impurity atom has only three electrons. All three electrons form bonds with surrounding atoms, but a "hole" remains in the crystal where the fourth electron should be, as shown in *Figure 8-2B*. An electron from elsewhere in the crystal can move to fill the hole, which in effect causes the hole to move to the place where the electron used to be. Thus, the hole can move or carry current in the semiconductor. Because the hole is an absence of a negatively charged electron, it is said to have a positive charge, and materials doped with electron acceptors that have three electrons in their outer shells are called *p-type* semiconductors.

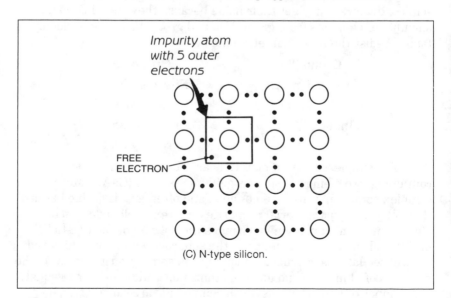

(C) N-type silicon.

and doped silicon crystals.

Elements such as aluminum or gallium are electron acceptors in silicon.

To complete the picture, we should point out that pure semiconductors are called *intrinsic* materials, because their natural conductivity is intrinsic to the semiconductor. Intrinsic semiconductor materials are sometimes are called *i-type* materials, but they aren't used often.

In practice, n- or p-type semiconductors are much more conductive than intrinsic materials, because the impurities provide many more current carriers. The degree of conductivity depends on the amount of doping. Impurity levels and conductivity can differ between different n- or p-type materials, even in different layers of the same device.

Compound Semiconductors

Silicon and germanium were the first semiconductors discovered, but there are many others. Not all of them are nominally pure elements. Certain compounds have nearly the same crystal structure and also function as semiconductors.

The most important of these are the so-called III-V semiconductors. They get their name because they contain equal amounts of elements from group IIIa and group Va of the periodic table. We list the important elements below:

Group IIIa	Group Va
Aluminum (Al)	Nitrogen (N)
Gallium (Ga)	Phosphorous (P)
Indium (In)	Arsenic (As)
	Antimony (Sb)

The simplest of these materials are "binary" compounds containing two elements, such as gallium arsenide (GaAs). More complex compounds also are possible, and often are desirable because they let you adjust material properties. For example, addition of aluminum to gallium arsenide changes the spacing of the crystalline lattice and the energy bandgap of the semiconductor (something that we will see later is of crucial importance in semiconductor lasers). The numbers of aluminum and gallium atoms must add together to equal the number of arsenic atoms. Such compounds are called "ternary" because they contain three elements, and are written in the form:

$$Ga_{1-x}Al_xAs$$

where,
   x is a number between 0 and 1.

This format indicates what we said before, that the number of gallium atoms plus the number of aluminum atoms must equal the number of arsenic atoms.

   You can get even more flexibility and control over material properties by adding a fourth element to the brew to make a "quaternary" compound. This is something like

$$In_{1-x}Ga_xAs_{1-y}P_y$$

where,
   x and y are numbers between 0 and 1.

In this case, the total number of indium and gallium atoms must equal the number of arsenic and phosphorous atoms.

   The more elements you add to the mixture, the harder it gets to grow good crystals. Other complications also arise, because the different materials don't always dissolve well in each other. However, ternary and quaternary compounds do have important attractions. By continuously varying the concentrations of gallium and aluminum, for example, you can vary the bandgap and lattice spacing of GaAlAs between those of GaAs and AlAs. Thus you can make a variety of compounds with different properties.

   Several other families of compounds, like the III-V compounds, form semiconducting solids. Some of them are useful for optical detectors. The only family useful in semiconductor lasers are the "lead salt" compounds which contain elements from columns IIB, IVB and VI of the periodic table. (They get their name because most of them contain lead.) The most important constituents of these lasers are:

Column IIB	Column IVB	Column VI
Cadmium	Tin	Sulfur
	Lead	Selenium
		Tellurium

Semiconductor Properties

We can break semiconductor properties into two broad categories: electronic and optical. Electronic properties include concentration of current carriers, conductivity, and mobility of electrons. These are important in the operation of all electronic semiconductor components, including lasers.

In semiconductor lasers (and in other optical devices), optical properties also are important. The key parameter behind most of them is the bandgap, the energy spacing between the conduction band and the valence band.

The bandgap gets its name because there are no energy levels between the conduction band and the valence band in semiconductors. If an electron is at the bottom of the conduction band, it must drop all the way to the top of the valence band. Likewise, an electron at the top of the valence band must jump the entire bandgap to reach the conduction band.

The implications this has for optical properties of semiconductors are shown in *Figure 8-3*. Let's first define a wavelength $\lambda_{bandgap}$ at which the photon energy $h\nu_{bandgap} = E_{bandgap}$. At longer wavelengths, the photon energy is smaller than the bandgap energy. Because photons with wavelengths greater than $\lambda_{bandgap}$ don't have enough energy to raise valence electrons to the conduction level, the valence electrons don't absorb them. Thus, the semiconductor is transparent to wavelengths longer than $\lambda_{bandgap}$. On the other hand, light at wavelengths shorter than $\lambda_{bandgap}$ has more than enough energy to raise valence electrons to the conduction band, so the semiconductor readily absorbs it. If the energy $E_{bandgap}$ is measured in electronvolts, its relationship to bandgap wavelength $\lambda_{bandgap}$ (in nanometers) is:

$$E_{bandgap} = 1240 \:/\: \lambda_{bandgap}$$

One note of caution: this picture is somewhat oversimplified. Semiconductors do not switch abruptly from complete absorption to complete transparency at $\lambda_{bandgap}$, but the change is fairly sharp.

Something similar happens to the light-emitting properties of semiconductors. If an electron in the conduction band drops back into the valence band, it can release the energy difference—at least equal to the bandgap energy—as a photon of light. Again, there are real-

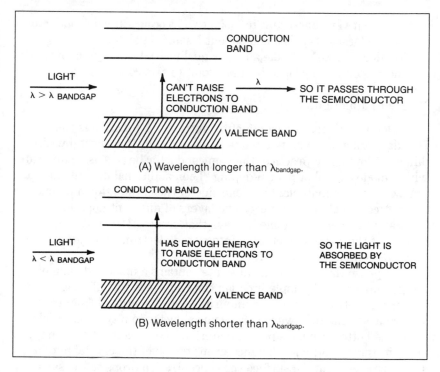

CONDUCTION BAND

LIGHT

$\lambda > \lambda$ BANDGAP

CAN'T RAISE ELECTRONS TO CONDUCTION BAND

λ

SO IT PASSES THROUGH THE SEMICONDUCTOR

VALENCE BAND

(A) Wavelength longer than $\lambda_{bandgap}$.

CONDUCTION BAND

LIGHT

$\lambda < \lambda$ BANDGAP

HAS ENOUGH ENERGY TO RAISE ELECTRONS TO CONDUCTION BAND

SO THE LIGHT IS ABSORBED BY THE SEMICONDUCTOR

VALENCE BAND

(B) Wavelength shorter than $\lambda_{bandgap}$.

Figure 8-3. How bandgap energy determines wavelengths at which semiconductors absorb, transmit, and emit light.

world complications to this picture, but the basic point is that the emission wavelength, too, is proportional to the bandgap energy.

The notion of bandgap does not apply only to semiconductors. We should note here that insulators also have bandgaps—but bandgaps too large for electrons to cross under normal circumstances. In conductors, in contrast, the bandgap is essentially zero, because there is no gap between valence and conduction electrons.

LIGHT EMISSION AT JUNCTIONS

While the bulk properties of semiconductors are important in determining how they will function in electronic devices, the "action" in a semiconductor usually takes place at the junction between two zones of the semiconductor with dissimilar impurity doping. The same is true for light emission, which is generated at the junction zone itself.

This makes it very important to look at what occurs at semiconductor junctions. Because this is a book about lasers, not about semiconductors, we will take a rather qualitative look at junctions, considering only the properties important to semiconductor lasers.

Nature of Junctions

Doping creates semiconductor junctions. A junction is the boundary zone between two parts of a semiconductor crystal with different doping. In practice, the semiconductor begins as a substrate with a modest level of p- or n-doping. Then additional dopants are diffused into the material from one side of the crystal, forming one (or in practice usually more) successive layers of different doping. Typically this doping is done in ways that create patterns on the surface of the semiconductor. In integrated electronic circuits, those patterns contain the circuit functions.

Suppose that the initial substrate contains a modest doping of p-type impurities. A simple junction can be formed by diffusing a higher concentration of n-type impurities into the crystal from the top. This converts the top layer of the semiconductor to n-type material, while the bottom remains p-type material. The concentration of n-type impurities drops off with distance from the top of the crystal until at some distance from the surface the concentration drops to the same level as the p-type impurities, as shown in *Figure 8-4*. That level is considered the junction between the n- and p-type materials. In practice, that level is thin by human standards (0.1 to 1 micrometer thick), but it represents many atomic layers. The hole concentration is roughly that of the p-type impurities; the electron concentration is roughly that of the n-type impurities.

The crucial variable in this junction device is the type of current carrier. This varies depending on whether or not voltage (or bias) is applied across the junction and—if so—the direction in which the voltage is applied. We will describe each of the three possible cases in qualitative terms.

Unbiased Junctions

If there is no bias across the junction, charge carriers are distributed through the crystal in roughly the same way as impurities. The electron carriers are common throughout the n region, while holes are common in the p region. Near the junction,

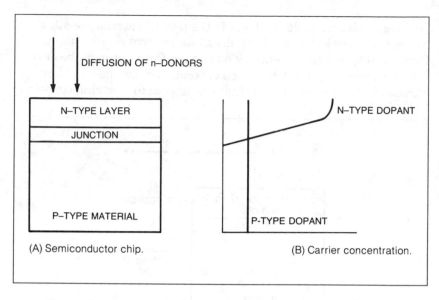

(A) Semiconductor chip.

(B) Carrier concentration.

Figure 8-4. Doping of a semiconductor crystal to form a junction.

the two types of carriers are present in roughly equal concentrations, and can cancel each other out by a process called "recombination."

Remember that a hole really is the absence of an electron in the valence band. This hole moves about the crystal when valence electrons shift their positions, creating a new hole while filling an existing one. However, a hole also can be filled by capturing a "loose" electron from the conduction band. The electron from the conduction band must lose energy when it drops to the valence band. If there is no electrical bias across the junction, this recombination on the average is balanced by the creation of new electron-hole pairs, and there is no net flow of current carriers in the crystal. Individual electrons and holes form, move, and recombine, but no current flows. The same thing happens in still air, where individual atoms and molecules move, but the gas as a whole stays in the same place.

Reverse-Biased Junctions

Now suppose that a positive voltage is applied to the n side of a junction and a negative voltage applied to the p side. As shown in *Figure 8-5*, the electrons in the n-type material are attracted to the

positive electrode, while the holes in the p-type material are attracted
to the negative electrode. This pulls all the current carriers away
from the junction and essentially no current flows through the device.
(There is, however, a small leakage current, because the
semiconductor does not have infinite resistance to electrical current.)

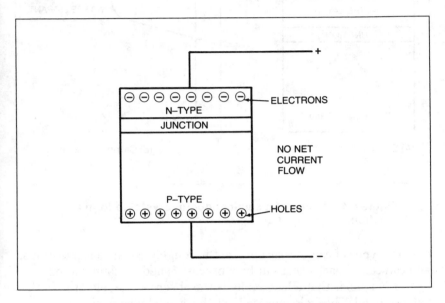

Figure 8-5. A reverse-biased semiconductor junction.

You will recognize this "reverse biasing" if you are familiar with
semiconductor electronics. This is the state in which a diode does not
conduct current, as long as the voltage is kept below a breakdown
threshold.

Forward-Biased Junctions

A semiconductor diode is said to be forward-biased if a positive
voltage is applied to the p side and a negative voltage to the n side.
As shown in *Figure 8-6*, this attracts the p and n carriers to the
opposite sides of the device, making them cross the junction. A
strong current starts to flow once the voltage passes the bandgap
energy measured in electronvolts, typically 0.5 to 2 electronvolts.

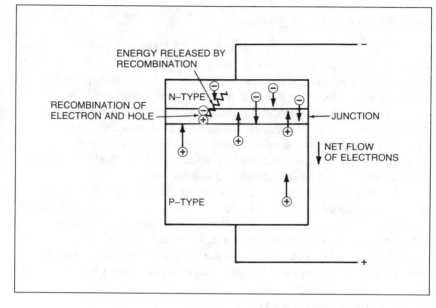

Figure 8-6. Forward-biasing triggers flow of electrons in a semiconductor diode.

In an electronic diode, the important thing is that current starts flowing when it is forward biased. In optical devices, something else important also happens at the junction. Electrons from the n-type material recombine with holes from the p-type material, releasing energy at the junction. (This energy is lost by the electron as it drops from the conduction to the valence band, and thus equals the bandgap. This bandgap energy accounts for the voltage drop across the forward-biased junction.)

In some semiconductors, such as silicon, the recombination energy is released as heat, warming the device. In others, such as gallium arsenide, much of the recombination energy is released as light. (The difference depends on details of semiconductor physics that you don't want to worry about.) Light emission from recombining electrons and holes can serve as the basis of light-emitting diodes (LEDs) and semiconductor lasers. The process that produces recombination is also called current injection, so semiconductor lasers sometimes are called "injection lasers" as well as "diode lasers" or "laser diodes."

8

<hr>

Optical Excitation at Junctions

Before moving on to the workings of LEDs and diode lasers, we should mention another process that occurs in semiconductor diodes: optical excitation. If a photon with more than the bandgap energy strikes a semiconductor, it can raise an electron from the valence band to the conduction band, creating an electron and a hole—a pair of current carriers. This effect can be used to detect light, by generating an electric current proportional to the amount of illumination, or by changing the semiconductor's conductivity, again by an amount proportional to the light intensity.

Light-sensitive semiconductors contain internal junctions, and are called "photodiodes." They are the most common type of detector used with lasers, and especially with semiconductor lasers. We mentioned them briefly in Chapter 5. We won't go into their operation here, but you should realize that semiconductor devices play a vital role in converting optical signals back into electronic form.

Light-Emitting Diodes (LEDs)

A light-emitting diode is a forward-biased semiconductor diode in which recombination at the junction produces light. As we mentioned earlier, only certain materials are suitable for LEDs. A sampling of these are listed in *Table 8-1*. Gallium arsenide doped with silicon is by far the most efficient, but its output is in the infrared.

LEDs are not lasers. The light they produce is spontaneous emission, just like that from a light bulb. Like a light bulb, an LED junction radiates light in every direction, as shown in *Figure 8-7*. To get the most efficient output, the junction should be close to the surface of the device. This reduces the chance that the emitted light will be absorbed by other parts of the semiconductor. (That absorption is possible because the material can absorb light on the same transition that it emits on.)

The structure shown in *Figure 8-7* is typical for LEDs intended to serve as small lamps, indicators, or displays. The LED may be packaged with a lens that concentrates the output emerging from the top surface. Other structures can concentrate the output into smaller areas, for applications such as coupling the light into an optical fiber. However, LEDs do not generate well-focused beams like lasers, and their output is spontaneous rather than stimulated emission.

Table 8-1. Some Important Commercial LED materials

Material	Dopant	Peak Wavelength or Range (nm)
GaP	Nitrogen	550–590 (green-yellow)
$GaAs_{0.15}P_{0.85}$	Nitrogen	589 (yellow)
$GaAs_{0.35}P_{0.65}$	Nitrogen	632 (orange)
$GaAs_{0.6}P_{0.4}$		650 (red)
GaP	Zn,O	700 (red)
GaAs	Zn	900 (infrared)
GaAs	Si	910–1020 (infrared)

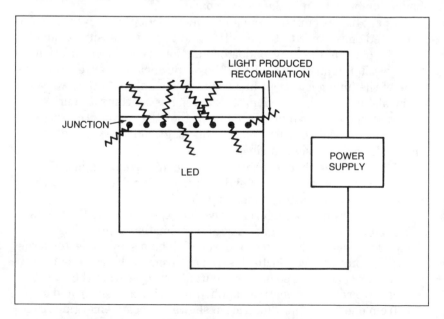

Figure 8-7. An LED emits light in all directions from the junction.

8

Nature of Semiconductor Lasers

Semiconductor lasers are closely related to LEDs. Both devices generate light from recombination of electron-hole pairs at a reverse-biased junction. In both, the light output is proportional to the drive current, and the output wavelength depends on the material's bandgap. Semiconductor lasers can produce low levels of incoherent emission—functioning like LEDs—when the current passing through them is below a threshold for laser action. However, there also are many important differences.

The first important difference is in current level. A modest current flows through an LED, causing some recombination that generates light. The current flowing through a diode laser is much higher, sufficient to produce a large concentration of recombining electron-hole pairs at the junction. The current concentration must be high enough for emission from the electron-hole pairs to dominate over absorption at the junction. That is, there must be a population inversion at the junction, with more electron-hole pairs in the conduction band than in the valence band.

Spontaneous emission can go in any direction. However, if stimulated emission is to be amplified and gain in intensity, it can only go in the plane of the semiconductor junction. Thus, laser emission is strongest in the junction plane, so the output beam emerges from the plane of the junction (except in a very few special cases that we won't worry about here). In practice, the emission is concentrated in a narrow stripe as shown in *Figure 8-8*, rather than across the entire junction plane. As we will see later, this diagram is very much simpler than actual semiconductor lasers.

The feedback that concentrates stimulated emission in the junction plane comes from the ends of the semiconductor crystal, which are smooth "facets" formed by cleaving the chip. Semiconductors have a high refractive index, so typically the facets reflect about 30% of the light back into the material, providing enough feedback for laser action in the high-gain semiconductor laser material. Some semiconductor lasers emit from both facets, but they normally are packaged so only one output emerges from the case. Others have reflective coatings applied to their rear facets, and only emit from one facet. Our illustration shows the beam emerging from only one end of the laser.

At low current levels, diode lasers generate some spontaneous

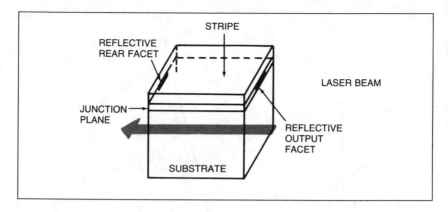

Figure 8-8. Light emission in a semiconductor laser is concentrated in the junction plane by feedback from the crystal's cleaved facets.

emission by the same processes that drive LEDs. However, as the current level increases, diode lasers pass a threshold where the population becomes inverted and laser action begins. This laser threshold is most visible in plots such as *Figure 8-9*, which shows output light power as a function of drive current. Below threshold, light is emitted with low efficiency. Once the current has passed the threshold (which is a characteristic of the individual laser), the light output rises steeply, showing the presence of stimulated emission.

Threshold current is a very important parameter in semiconductor lasers. Below threshold, the light output is very inefficient, and most energy in the drive current is lost as heat. Light emission becomes more efficient above threshold—meaning more of the input electrical energy emerges as light energy. The higher the laser threshold, the more electrical power is lost as heat which must be dissipated before you get any laser output. Because dissipating heat tends to shorten lifetime of a semiconductor laser, this means that lifetime declines as threshold current increases.

We should note that the figure of merit for semiconductor laser lifetime often is measured as threshold current density rather than the absolute value of threshold current. To get the density of the threshold current, we divide the current by the active area of the laser stripe to get amperes per square centimeter flowing through the laser junction. This current density measures the stress on the laser material.

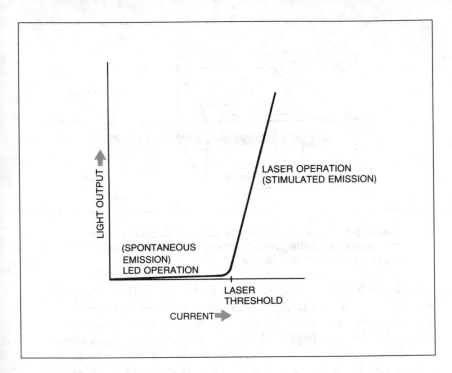

Figure 8-9. Light output of a semiconductor laser above and below laser threshold current.

Direct vs Indirect Bandgap Materials

All semiconductor laser materials can operate as LEDs, but many LEDs cannot operate as semiconductor lasers. The reason lies in the complexities of semiconductor physics that determine the nature of the energy levels. In what are called "direct bandgap" materials, all the energy released a transition from the conduction band to the valence band can be in the form of light. However, in what are called "indirect bandgap" materials, some transition energy must be released as vibrations of the crystal lattice. This makes it harder to produce light by recombination of current carriers in indirect-bandgap materials.

Heavily doped indirect bandgap materials can work as LEDs, but not as lasers. Unfortunately, most of the large-bandgap materials needed to produce visible light have indirect bandgaps, and thus are unsuitable for use as lasers. As we will see later, this materials problem is the major reason why semiconductor lasers have so far made little progress in invading the visible part of the spectrum.

STRUCTURES OF SEMICONDUCTOR LASERS

As we warned earlier, *Figure 8-8* gave a very oversimplified view of semiconductor lasers. The simple structure shown in that illustration was used in the first diode lasers, but developers who wanted to produce continuous-wave output at room temperature had to turn to much more complex structures to increase efficiencies and lifetimes and reduce threshold currents.

A diode laser must contain at least three layers—a p-type layer, an n-type laser, and a separate active layer (either n or p type) which emits the light. High-performance diode lasers usually contain more layers. All this complexity comes in a tiny package. Typical diode laser chips are 300 to 500 micrometers long, a fraction of that distance in width, and perhaps 10 micrometers thick. Bare laser chips sometimes are used in research laboratories, but for most applications they are packaged in small cases, many of which look like the metal cases used with old discrete transistors.

Homostructure and Heterostructure

The boundaries between the active layer and the surrounding layers are very important in determining efficiency of a diode laser. In the first semiconductor lasers, these layers were the same material, and because they were made of the same material, they formed what are called "homostructure" or "homojunction" lasers. Later developers turned to what are called "heterostructure" or "heterojunction" lasers, in which adjacent layers are made of different materials. The advantage of using different materials is that they help confine light more efficiently in the active layer. (Note that "homojunctions" and "heterojunctions" are not exactly the same as the simple p-n junctions we have been describing throughout this chapter, but represent boundaries between different layers in a multilayer semiconductor structure.)

Figure 8-10 shows the difference between homostructure and heterostructure lasers. In a homostructure laser *Figure 8-10A*, the semiconductor has essentially uniform refractive index throughout, so light building up in the active layer can diffuse into surrounding layers. However, the layers in a heterostructure laser, have different

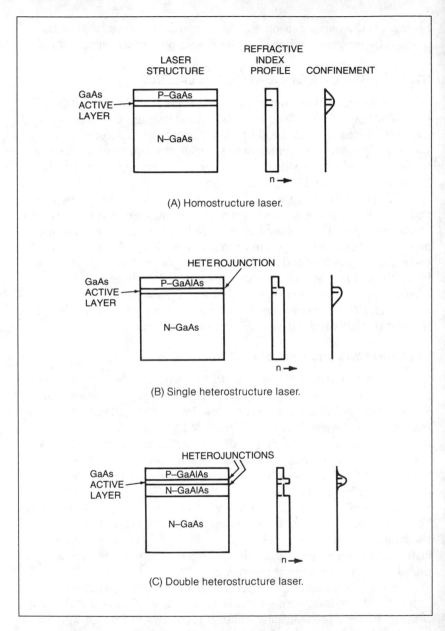

(A) Homostructure laser.

(B) Single heterostructure laser.

(C) Double heterostructure laser.

Figure 8-10. Homostructure and heterostructure lasers, with refractive index profile and light confinement.

compositions and thus different refractive indexes. In a double-heterostructure laser *Figure 8-10C*, the active layer has a higher refractive index than either of the two adjacent layers. This helps confine light generated in the active layer by creating a waveguide layer (much like the refractive-index difference between the core and cladding of an optical fiber confines light in the core). In a single-heterostructure laser *Figure 8-10B*, the refractive index changes significantly on only one side of the active layer. This gives better confinement than a homojunction laser, but not as good confinement as a double heterostructure laser.

The differences in light confinement lead to important differences in functional characteristics. Homostructure lasers let so much light leak out of the active layer that they are too inefficient for practical use. Single-heterostructure lasers are more efficient, but they do not confine light well enough to operate continuous wave at room temperature. Their main uses are to generate pulses lasting under a microsecond with peak powers of one to a few tens of watts at low repetition rates.

Double-heterostructure lasers are much more efficient, and can generate a continuous-wave beam at room temperature. That approach is used in most commercial diode lasers, and has been refined in many ways.

Stripe-Geometry Lasers

We mentioned earlier that the light-emitting areas of diode lasers normally are stripes rather than the entire width of the active layer. Stripe-geometry lasers have much better beam quality than broad-area lasers. Typically the stripes are only a few micrometers wide. They can be defined in two ways: by boundaries where laser gain drops off, or by changes in refractive index of the material itself.

The basic concept of a gain-guided laser is shown in *Figure 8-11A*. An insulating layer is grown on top of the laser chip, blocking current flow at the sides, and confining it to a narrow stripe the length of the chip. Only in that narrow stripe does enough current flow to produce a population inversion and the right conditions for laser gain. Because there is no gain at the sides, there is no emission from those regions, even though there is no physical boundary separating the stripe from the rest of the active layer.

Index-guided lasers are an extension of the concept behind

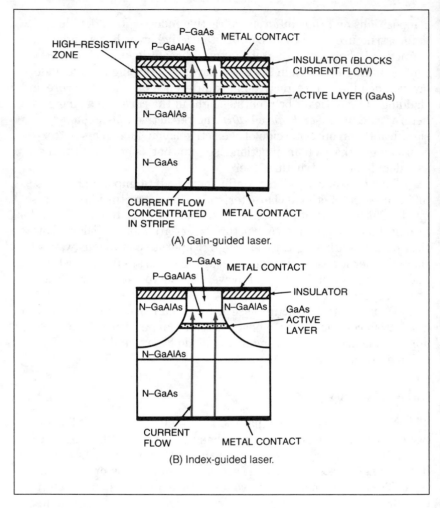

Figure 8-11. Gain-guided and index-guided lasers.

double-heterostructure lasers. In such a laser, the stripe is defined by a change in refractive index where the composition of the laser material changes. In the example shown in *Figure 8-11B*, the current flows through the central "mesa" to the active-layer stripe buried below. An insulator layer on the top of the laser prevents current from flowing off to the sides, but the composition also changes at the sides of the stripe, confining laser light in much the same way as the waveguide formed in a double-heterostructure laser. Lasers in which the entire stripe is entirely "buried" by other layers—as in *Figure 8-11*—sometimes are called "buried heterostructure" lasers.

Laser Arrays

In our examples so far, each laser chip contains only one laser stripe. However, in the past few years developers have made chips containing many parallel stripes. These monolithic laser arrays can produce much more power than single laser stripes, although it is hard for them to match the beam quality of single laser stripes.

A typical single-stripe laser can generate continuous-wave powers from a few milliwatts to tens of milliwatts. Arrays with tens or even hundreds of parallel laser stripes can generate continuous-wave powers of hundreds of milliwatts to a watt or more. Even higher powers are possible if the laser is operated in the repetitively pulsed mode.

Monolithic laser arrays have reached the market only in the past few years, and the technology continues to develop rapidly. Because the output comes from many laser stripes, the beam quality is not as good as the best single lasers. Progress is being made on improving beam quality, but in the meanwhile high-power diode-laser arrays are opening major new vistas for semiconductor lasers in applications that require powers higher than previously available.

Distributed Feedback Lasers

Some applications require semiconductor lasers that emit only a narrow range of wavelengths—a stable single longitudinal mode. This is difficult with conventional diode lasers. The short cavity length of diode lasers does limit the number of longitudinal modes possible, because there are few values of N that solve the equation

$$2L = N\lambda$$

when the cavity length, L, is only 500 micrometers long. For a typical gallium-arsenide laser emitting near 900 nm, it means that adjacent longitudinal modes are about 1 nm apart.

Semiconductor lasers have a gain bandwidth broad enough to span several longitudinal modes. Some—but not all—side modes are suppressed by the processes that amplify laser emission, and typically one mode dominates. Unfortunately, changes in temperature and other operating conditions can cause a diode laser to "hop" from one dominant mode to an adjacent one. This leads to undesirable instability in laser operation and output wavelength. The shifts in wavelength can be particularly harmful in certain fiber-optic systems operating near the 1.55-micrometer wavelength of minimum fiber loss.

The cure is a diode-laser design that includes mechanisms to stabilize the laser's output wavelength. Currently a leading approach is the distributed feedback laser, shown in *Figure 8-12*. It contains a diffraction grating that scatters light back into the active layer. The grating can be along the length of the part of the active layer where there is gain *(Figure 8-12A)*, or can extend beyond the ends of the region where there is gain *(Figure 8-12B)*. The feedback leads to

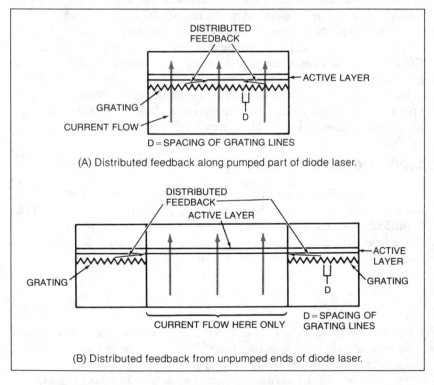

(A) Distributed feedback along pumped part of diode laser.

(B) Distributed feedback from unpumped ends of diode laser.

Figure 8-12. Two types of distributed-feedback lasers.

interference effects which allow oscillation only at wavelengths which are reinforced by the scattering effects.

The grating spacing (D) needed to select a particular wavelength (λ) in a distributed-feedback laser depends on the refractive index in the laser media n and a positive integer m that gives the order of distributed-feedback coupling

$$D = m\lambda \,/\, 2n$$

Typical values of m are 1 or 2. If we plug in the appropriate numbers for a 1.55-μm distributed-feedback laser made of InGaAsP ($n = 3.4$), we find that D is 228 nm for $m = 1$ and 456 nm for $m = 2$. Thus the distributed feedback limits laser oscillation to a single longitudinal mode, giving the laser a very narrow bandwidth. The oscillation wavelength does shift slightly with temperature because the refractive index n is a function of temperature, but this change is much smaller than the wavelength shifts in conventional diode lasers.

The distributed feedback laser has been the most successful type of "single-frequency" diode laser, and is the only type that has been offered commercially. Two other types have attracted considerable attention, but remain in the research stage. One is the external-cavity laser, in which external optics extend the cavity of a diode laser beyond the semiconductor chip itself and limits laser oscillation to a narrow range of wavelengths. The other is the cleaved-coupled-cavity (C^3) laser, in which a single laser chip is split so the two halves can be coupled optically but driven separately electronically. The optical coupling limits oscillation to a narrow bandwidth.

Superlattice Structures

Semiconductor lasers contain layers as thin as a hundred nanometers. Recently, researchers have found that they can build interesting structures by successively depositing many layers that are each only a few nanometers thick. Typically alternating layers have different compositions—for example, one might be GaAs, the other GaAlAs. One group of such devices is called "quantum well" devices because the differences in bandgap and energy levels between the layers create leaky quantum-mechanical traps for electrons and holes. Another is called "superlattices," because of the many layers they contain.

Superlattice structures may help semiconductor physicists overcome one of their biggest practical material problems: the need for lattice matching. Now, you can grow one semiconductor material on top of another only if their interatomic spacings are very closely matched. Otherwise, the atoms will not mesh properly and the crystal will be fatally flawed. Structures called strained-layer superlattices can match two materials with different composition by gradually grading the lattice spacing of a series of many thin layers. The thin layers can better withstand strain caused by the difference in atomic spacing than the thicker layers in conventional semiconductor devices.

Some specialists hope that development of superlattices and other ultrathin-layer semiconductor structures will let them custom-tailor the properties of materials. One major goal is materials that could provide visible-wavelength semiconductor lasers.

BEAM CHARACTERISTICS AND STRUCTURE

As with other lasers, the characteristics of diode-laser beams depends heavily on their structure and lasing characteristics. These differ markedly from those of other types of lasers.

Diode laser materials have high gain, and their cavities are very short—typically 300 to 500 micrometers long. This means that they are only several hundred wavelengths long, compared to hundreds of thousands of wavelengths long for a typical visible-wavelength gas laser. The end mirrors on the cavity are the flat, cleaved ends (called facets) of the semiconductor crystal. The rear facet may (or may not) have a reflective coating; the front facet generally does not. The materials used in semiconductor lasers have high refractive index; for example, the refractive index of gallium arsenide is 3.34 at 780 nm. This causes high reflection at the semiconductor-air interface, so about 30% of the light is reflected back into the semiconductor. This provides adequate feedback for laser oscillation. However, as in other high-gain lasers, the cavity does not limit oscillation to a TEM_{00} mode.

Beam Size and Divergence

We saw earlier that beam divergence depends on the nature of the resonant cavity and the size of a laser's emitting area. In diode lasers, the small cavity combines with an unusually shaped emitting area to produce a noncircular beam that spreads unusually rapidly.

The shape of a diode-laser beam depends mainly on its emitting area. We have assumed that gas and solid-state lasers emit round beams, and for most purposes that is a good assumption. Diode laser beams are not round because the active layers are not circular. Typically, active layers are about a fraction of a micrometer thick and several micrometers across. (These dimensions are difficult to show to scale, so we haven't tried in our illustrations.) At the facet ends, they look like long, thin rectangles, but the beams they produce are smoother around the edges, and are long, thin ovals.

Earlier, we saw that beam divergence roughly equals the wavelength divided by the output-beam diameter. For noncircular beams, the divergence can be different in the two perpendicular directions. It is larger in the direction of the smaller dimension (active-layer thickness) and smaller in the direction of greater active-layer width.

For a gas laser, the beam diameter at the output mirror typically is over a thousand times the wavelength, but for a semiconductor laser the ratio is much smaller. For a typical 800-nm GaAlAs/GaAs laser, the light-emitting stripe is about 10-wavelengths wide and only about a quarter of a wavelength high. These values are large enough that the simple approximations we used to calculate divergence break down. It is easier to give the measured values for beam divergence: 10 degrees (0.17 radian) in the direction parallel to the active layer and 40 degrees (0.70 radian) in the direction perpendicular to the active layer. Compare those to the typical 0.001 radian divergence of a helium-neon laser, and you realize you have a rather different beast. The beam from a diode laser spreads out more rapidly than one from a good flashlight!

Fortunately, it is possible to add external optics to correct for this broad beam divergence. A cylindrical lens, which refracts light in one direction but not in the perpendicular direction, can make the beam circular in shape. Collimating lenses can focus the rapidly diverging beam to a much narrower divergence, like a flashlight lens focuses light from a flashlight bulb into a beam.

Modes and Coherence

Narrow-stripe diode lasers can oscillate in a single transverse mode, and are often called "single-mode" lasers. However, as with gas lasers, that single transverse mode can include more than one longitudinal mode. As long as a diode laser emits multiple longitudinal modes, its bandwidth remains large enough to severely limit coherence length. Limiting oscillation to a single longitudinal mode can give coherence lengths of tens of meters, but it requires special structures like the distributed-feedback laser described earlier.

The poor coherence length of standard semiconductor lasers sets them apart from gas lasers. Unlike their broad beam divergence, it cannot be corrected just by adding simple external optics. This is a serious problem for holography, and a few other applications which require coherent light. However, coherence is unnecessary for most applications, and in some cases can introduce extraneous noise in the form of speckle, as described in Chapter 4. Thus the short coherence of semiconductor lasers is not always bad news.

Modulation

The output of a diode laser depends upon the drive current flowing through it. Once the drive current has passed laser threshold, the light output increases roughly linearly with current, as shown in *Figure 8-9*. (The light-current curves of some diode lasers have "kinks" that mark sudden changes in light emission or current level, but for many purposes lasers with such kinks are considered defective.)

This simple dependence of light output on drive current makes direct modulation merely a matter of changing the drive current. The laser emission responds quickly, especially when the light-emitting stripe is narrow, and current flow is confined to a small area. The best diode lasers have bandwidths in the gigahertz range (billions of hertz), although not all diode lasers can respond that rapidly.

Some problems can occur at very high speeds. One that may limit the speed of direct modulation for fiber-optic communications is a change in wavelength that accompanies direct modulation. The change in electron density during the current pulse changes the semiconductor's refractive index n. This enters into the equation for the laser oscillation wavelength λ (measured in air) given by the requirement that the round-trip length of the cavity $2L$ be equal an integral number of wavelengths in the laser medium.

$$\lambda = 2\,Ln\,/\,N$$

The change is not large, but because the refractive index of optical fibers varies with wavelength, pulses containing a range of wavelengths spread out as they travel through the fiber. If the spreading is too large, the pulses will interfere with each other, limiting transmission speed. To avoid this problem, developers of ultrahigh-speed fiber-optic systems are working on external modulators which modulate the beams from continuous-wave diode lasers without changing their wavelengths.

WAVELENGTHS AND MATERIALS

Earlier, we indicated that a diode laser's wavelength depended on the size of the material's bandgap. The photon energy does not exactly equal the bandgap energy, but it is close. If bandgap energy E is given in electronvolts (as is common), you can quickly calculate the approximate emission wavelength λ (in nanometers) from the formula:

$$\lambda = 1240 / E$$

The bandgap energy in electronvolts also is close to the voltage drop across a forward-biased diode.

The bandgap energy, and thus the emission wavelength, depend on lattice spacing and semiconductor composition. Both lattice spacing and bandgap energy are fixed for a pure binary semiconductor such as gallium arsenide, indium phosphide, or lead sulfide. However, you can get other wavelengths by forming semiconductor crystals from mixtures of three or four elements. You can think of the process as blending different materials to get intermediate characteristics. For example, a mixture of 20% AlAs and 80% GaAs has bandgap energy and lattice spacing between pure AlAs and pure GaAs.

Things get more complex if the compound semiconductor contains four elements. Depending on the relative concentrations of the elements, such "quarternary" semiconductors can have properties somewhere in a broad range. For example, as shown in *Figure 8-13*, InGaAsP (more precisely $In_xGa_{1-x}As_{1-y}P_y$) can have lattice spacing and bandgap within an area defined by four possible binary compounds: InAs, InP, GaAs, and GaP. The material's properties can vary within the entire space because the In-Ga and As-P ratios can be adjusted independently. (The odd shape of the area in the drawing arises because pure GaP is an indirect bandgap material, unsuitable for lasers and somewhat different from the other binary

compounds. The dashed line indicates the indirect bandgap region.
The line between GaAs and AlAs indicates the range for GaAlAs.)

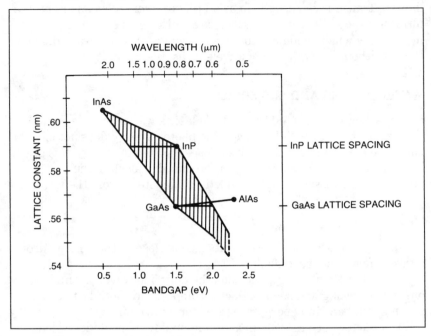

**Figure 8-13. Lattice constant vs bandgap for InGaAsP can
fall within the shaded area defined by properties of InAs,
InP, GaAs, and GaP.**

As we mentioned earlier, lattice spacing also is a crucial variable
for semiconductor laser materials. To avoid flaws that degrade device
performance, the entire structure must be made of compounds with
nearly identical lattice spacing. This means that all compositions in an
InGaAsP structure must fall roughly on a horizontal line in *Figure 8-13*.

In practice, the lattice spacing is set by the choice of substrate
material. Binary compounds normally are used as substrates because
they are much easier to produce, and the substrate accounts for most
of the device volume. However, the choice of a binary compound
restricts the choice of other materials. If the substrate is GaAs, you
can only use mixtures with the same lattice spacing. Look closely at
the figure, and you will note that the only compounds that match
GaAs spacing are those with larger bandgaps and output wavelengths

shorter than the nominal 900 nm of GaAs. If you want longer wavelengths from InGaAsP compounds, you must use InP as a substrate.

The substrate is important enough that developers often indicate it in describing the structure of semiconductor lasers. The composition of the active layer is first, then that of the substrate. For example, an InGaAsP/InP laser has an active layer of InGaAsP and a substrate of InP. Other deposited layers may have other compositions, generally ternary (three-element) compounds that are lattice-matched to the substrate and active layer.

Developers have concentrated primarily on four families of semiconductor laser materials: GaAlAs, InGaAsP, AlGaInP, and lead-salt materials. The wavelength ranges of those and other developmental materials are shown in *Table 8-2*.

Table 8-2. Major Semiconductor Laser Materials

Material	Wavelength Range	Comments
GaInPAs/GaAs	600–800 nm?	Very hard to grow
AlGaInP/GaAs	580–690 nm	Very hard to grow
GaInP/GaAs*	670–680 nm	Easier to grow because no Al
GaAlAs/GaAs*	720–900 nm	Short lived at short wavelengths
GaAs/GaAs (pure)*	904 nm	
InGaAs*	1060 nm	
InGaAsP/InP*	1000–1700 nm	
PbCdS	2.7–4.2 μm	Requires cooling
PbSSe	4.2–8 μm	Requires cooling
PbSnTe	6.5–30 μm	Requires cooling
PbSnSe	8–30 μm	Requires cooling

*Capable of continuous-wave operation at room temperature over at least part of wavelength range

8

Gallium-Arsenide Lasers

The best-developed family of diode lasers are those made from GaAlAs and GaAs. They usually are called simply "gallium arsenide" lasers, because GaAs is the substrate material and the aluminum concentration normally is small.

The nominal wavelength of pure GaAs lasers is 904 nm, but shorter wavelengths are possible if aluminum is added to the mixture. However, adding aluminum beyond a certain point increases the likelihood of flaws that can quickly cause failure of the lasers. The shortest wavelengths readily available from commercial GaAlAs lasers are about 750 nm. Shorter-wavelength GaAlAs lasers have been made in the laboratory, but they are short-lived.

The near-infrared wavelengths of GaAlAs lasers are adequate for many purposes but ideal for few. They can be used in applications including fiber-optic communications, pumping of solid-state lasers, reading optical disks, and writing in laser printers. However, other wavelengths would be better for many applications. Optical fibers have lower attenuation and broader bandwidth at longer wavelengths, which led to the development of InGaAsP lasers. Shorter wavelengths would allow smaller focal spots, and would work better with many light-sensitive materials. In addition, shorter wavelengths would be visible, letting diode lasers replace helium-neon lasers for jobs that require visible light.

The strength of GaAlAs laser technology is that the material is well-developed and comparatively easy to use. These are far from trivial advantages because they permit production of both high-power and low-cost devices. All commercial high-power monolithic laser arrays are made of GaAlAs. The 800–900 nm output may not be ideal for all purposes, but it matches the absorption bands of neodymium lasers, and can be used for other applications where high power from a small laser is advantageous. Power levels are reaching levels where frequency doubling may become attractive, leading to generation of second-harmonic wavelengths of 400 to 450 nm.

Low manufacturing costs have been crucial to the development of consumer products using diode lasers. The most successful of these is the compact disc audio player, which uses a GaAlAs laser to play back digital sound from disks 12 centimeters (4 ¾ inches) in diameter. The least expensive players sell for under $100 each at discount stores. Some 10 million diode lasers were produced for optical disk players in 1987, and in large quantities they reportedly sell for only a few dollars each.

InGaAsP Lasers

The development of InGaAsP lasers was stimulated by the need for lasers with wavelengths long enough to take advantage of the low loss of optical fibers at wavelengths longer than 1 micrometer. As a quaternary system, the characteristics of $In_xGa_{1-x}As_yP_{1-y}$ can be varied continuously by varying x and y independently. When lattice-matched to an InP substrate, InGaAsP lasers can generate wavelengths of 1000 to 1700 nm. Standard designs use InGaAsP for the active layer, but other layers deposited on the InP substrate often are not the quarternary compound.

Most work has concentrated on two specific wavelengths—the 1300-nm wavelength where conventional single-mode optical fibers have their highest transmission bandwidth, and the 1550-nm wavelength where they have their lowest attenuation. Lasers emitting at 1300 nm were developed quickly, but commercial development of 1550-nm lasers took more time because materials problems caused low production yields.

The need to precisely control concentrations of four elements makes InGaAsP a difficult material to produce, making costs—and prices—much higher than for GaAlAs lasers. Prices have dropped as manufacturing techniques have improved, but InGaAsP lasers remain high-performance devices for applications that demand them. That is particularly true for 1550-nm lasers, which are much more costly than 1300-nm lasers.

Practical production limitations make the 1300- and 1550-nm wavelengths nominal values. The actual values of production lasers typically is specified only to within ± 20 or 30 nm because it is hard to control the composition precisely enough to make lasers with output at a specific wavelength.

Visible Diode Lasers

As this book is being readied for publication, developers finally succeeded in one of the longest-running quests in the laser business—the search for visible-wavelength diode lasers. The first commercial products were announced in early 1988. With active layers of GaInP, they emit red light at 670 and 680 nm, and are part of the larger family of AlGaInP semiconductor lasers that have large bandgaps and hence short emission wavelengths.

The GaInP lasers are not the first to be called "visible" diode lasers, although other claims are somewhat misleading. GaAlAs lasers can emit wavelengths of 720 to 750 nm that are faintly visible to the human eye, and some manufacturers have called such devices "red" lasers. However, that is a potentially dangerous misnomer. At wavelengths longer than the 633-nm helium-neon line, the human eye's sensitivity drops by a factor of 10 every 25–30 nm. Thus, at 750 nm, the eye is only about 0.0001 as sensitive as at 633 nm, while at 720 nm it is only 0.001 as sensitive. This means that if the beam from a helium-neon laser looks about as bright as that from a 720-nm diode laser, the latter contains about 10,000 times more power. (This deceptive appearance makes the beams from such very near infrared diode lasers potentially dangerous, because the viewer is unlikely to realize that they actually contain considerable power invisible to the human eye.)

It is more proper to consider 700 nm as the long-wavelength end of the visible spectrum, and developers have tried to reach that point for years. In fact, the passed it many times, but not for very long. Short-wavelength semiconductors lasers have been notoriously short lived. As we mentioned earlier, adding enough aluminum to make the wavelength shorter makes GaAlAs too prone to fatal flaws.

The new visible diode lasers have active layers of GaInP, and generate a few milliwatts at red wavelengths of 670 to 680 nm. Estimated lifetimes are around 10,000 hours—not as long as InGaAsP or GaAlAs lasers, but long enough for many applications. Initially offered in sample quantities at a few hundred dollars each, they are likely to drop in price as production increases. As prices drop, they will start to replace red helium-neon gas lasers in applications that do not require long coherence length, such as bar-code readers for supermarkets.

GaInP lasers are a subset of the larger family of AlGaInP lasers, in which the number of aluminum, gallium, and indium atoms together equal the number of phosphorous atoms. (You could write this $Al_yGa_xIn_{1-x-y}P$.) As in GaAlAs, addition of aluminum increases the band-gap energy, leading to shorter wavelengths. AlGaInP lasers can generate wavelengths between 580 and 690 nm. However, lifetimes remain limited at the shorter wavelengths, and only GaInP lasers are available commercially.

Lead-Salt Diode Lasers

As we mentioned earlier, lead-salt diode lasers are not members of the III-V compound semiconductor family that includes GaAlAs and InGaAsP lasers. They emit light at longer infrared wavelengths than III-V lasers, as shown in *Table 8-2*, require cooling to low temperatures (generally under 100 K or $-170\,°C$), and are used in specialized scientific applications.

Lead-salt lasers have a different crystalline structure than III-V semiconductors. They also have smaller bandgaps, under half an electronvolt. That small bandgap leads to emission at much longer wavelengths, beyond 2.7 micrometers. The size of the bandgap depends on the semiconductor composition, and like that of III-V compounds it can be varied continuously by adjusting the relative concentrations of two elements in a ternary compound. For example, changing the composition of $PbS_{1-x}Se_x$ from PbS (x = 0) to PbSe (x = 1) changes the emission wavelength from 4.2 micrometers to 8 micrometers. The longer the wavelength, the lower the temperature to which the laser must be cooled to operate properly.

The only practical use of lead-salt diode lasers is in infrared spectroscopy, the measurement of how various materials behave when illuminated with infrared light. Infrared measurements can be made with other light sources, but the diode laser emits a narrower range of wavelengths. The nominal wavelength of an individual laser is set when it is made, but the emitted wavelength can be adjusted somewhat around this value by changing the pressure, magnetic field, and temperature.

SPECIALIZATION OF DIODE LASERS

Semiconductor-laser technology has become specialized. Early diode lasers were made for general-purpose use and packaged in metal housings similar to standard transistor cases. Some general-purpose diode lasers are still made, but most are built and packaged with specific applications in mind.

Optical Disk Lasers

The biggest single use for any laser is the use of diode lasers to play compact disc digital audio recordings. GaAlAs diode lasers are mass-produced for this application. The lasers are mounted directly in a playback head that includes focusing optics and a light detector.

Because of the economies of scale in semiconductor production, identical diode-laser chips may be packaged in other ways for other applications where low cost is more important than matching laser characteristics specifically to the application. For example, audio disk lasers are packaged in light pens, and used in certain types of inexpensive fiber-optic transmitters.

Fiber-Optic Lasers

Both GaAlAs and InGaAsP diode lasers are used in fiber-optic systems. The short-wavelength GaAlAs lasers generally cost less, but fiber losses are higher at their operating wavelength, so they normally are used only in short-distance applications. Longer-wavelength but more expensive InGaAsP lasers are used when transmission distance is more important. For less-demanding applications where data rates are modest (typically 10 megabits per second or less) and transmission distances are short, LEDs of either GaAlAs or InGaAsP can be used in transmitters at significant cost savings.

Laser packaging is a prime consideration, and a major contributor to cost of transmitters for single-mode fiber systems. The laser's light-emitting area must be coupled to the light-carrying core of a single-mode fiber. The problem is that the fiber cores are only 8 to 10 micrometers in diameter and the lasers' light-emitting areas are even smaller, imposing tight tolerances on mechanical alignment. The laser can be packaged in a housing that mates with a fiber-optic connector, or in housing that includes a fiber-optic "pigtail" that can be spliced into the fiber-optic system.

Developers are working on ways to reduce costs of fiber-optic lasers by automating packaging. This may reduce performance as measured on some scales, such as output power emerging from the connector or fiber pigtail. However, developers hope that by reducing the cost they can increase the market greatly, and make it possible for fiber-optic connections to reach all the way to homes.

High-Power Lasers

Another fast-developing area of diode-laser technology is high-power monolithic arrays. As we mentioned earlier, the diode array is a large-area chip containing many parallel laser stripes. These lasers are all driven by the same source. Although individual stripes have been modulated separately in some laboratory versions, the prime emphasis now is on generating power, without worrying too much about modulation speed or beam quality. The major application of these laser arrays so far has been in pumping neodymium-YAG lasers, as mentioned in Chapter 7, for which neither beam quality nor modulation speed is critical.

Heat dissipation remains a concern for diode-laser arrays despite their high efficiency. Thus, such lasers are packaged with heat sinks to remove excess heat before it destroys the laser. This cooling requirement could impose a limit on power levels available from diode lasers.

8

What Have We Learned?

- Semiconductor lasers emit light when current flows through a junction of differently doped materials.

- The properties of semiconductors depend on the bonding and motion of electrons.

- n-type semiconductors are doped with elements that release electrons to carry current. p-type semiconductors are doped with elements that form holes as current carriers.

- III-V compounds such as gallium arsenide are called compound semiconductors. Their properties depend on the composition.

- Ternary semiconductors contain three elements, such as $Ga_{1-x}Al_xAs$. Quarternary semiconductors contain four elements, such as $In_{1-x}Ga_xAs_{1-y}P_y$.

- Lead-salt compounds also can be used in semiconductor lasers.

- Semiconductors absorb photons with energy greater than the bandgap energy, and transmit photons with less energy.

- Light emission wavelengths of semiconductors depend on bandgap energy.

- A free electron can fall into a "hole", eliminating both current carriers by a process called "recombination."

- If a positive bias is applied to the n material and a negative bias to the p material, no current flows across the junction.

- A semiconductor junction carries current if a positive voltage is applied to the p material and a negative voltage to the n material.

- A photon with energy greater than the bandgap can create an electron-hole carrier pair, an effect useful in detecting light.

- LEDs are forward-biased semiconductor diodes that generate light by recombination. Only some semiconductor materials work as LEDs.

- Semiconductor lasers produce stimulated emission above a threshold current level. They operate at higher currents than LEDs and concentrate emission in the junction plane.

- Stimulated emission in a semiconductor laser is concentrated in the junction plane by feedback from cleaved ends of the chip.

- Only direct-bandgap semiconductors can operate as lasers.
- Complex internal structures are needed for continuous-wave room-temperature diode lasers.
- Normally laser action is concentrated in a narrow stripe on the active layer.
- Arrays of many laser stripes in parallel on one chip can produce high power output.
- Distributed-feedback lasers have very narrow emission bandwidths.
- Extremely thin layers can be built up in superlattice and quantum-well structures. These are promising for future diode lasers.
- Diode lasers have high gain and very short cavities, making them different in important ways from other lasers.
- Diode laser beams are oval in cross-section and diverge very rapidly.
- The emitting stripe of an 800-nm GaAlAs laser is about 10 micrometers wide and a quarter-micrometer thick.
- Diode lasers have short coherence length unless they operate in a single longitudinal mode.
- Diode lasers are simple to modulate by changing their drive current.
- Bandgap energy and laser wavelength depend on the composition of the semiconductor, which can be adjusted in a three- or four-element compound semiconductor..
- GaAlAs is the best-developed and highest-power diode-laser material. Wavelengths are 750–900 nm.
- GaInP diode lasers can emit a few milliwatts at 670–680 nm.
- Lead-salt lasers emit at wavelengths longer than 2.7 μm and require cryogenic cooling. Their major uses are in research.
- Millions of diode lasers are made expressly for compact disc audio players.
- Diode lasers are packaged with fiber-optic connectors or pigtails for communication applications.
- High-power diode laser arrays are being developed for pumping Nd-YAG lasers and other applications.

8

WHAT'S NEXT?

In Chapter 9 we will talk about the types of lasers that didn't fall into our major categories: tunable organic-dye lasers, free-electron lasers, and X-ray lasers.

Quiz for Chapter 8

1. What type of semiconductor is doped with impurities that create holes as current carriers?
   a. intrinsic
   b. n-type
   c. p-type
   d. insulating
   e. None of the above

2. Which of the following is a quaternary III-V semiconductor?
   a. InGaAsP
   b. PbSnSSe
   c. GaAlAs
   d. GaAs
   e. NSbAsP

3. A semiconductor has a bandgap of 1.5 electronvolts. At about what wavelength will it emit light if it can operate as a laser?
   a. 1.5 μm
   b. 1000 nm
   c. 827 nm
   d. 678 nm
   e. 667 nm

4. What type of semiconductor junction can function as a laser?
   a. Unbiased junction
   b. Forward-biased junction
   c. Reverse-biased junction
   d. All of the above
   e. None of the above

5. How does a semiconductor laser operate when the drive current is below laser threshold?
   a. As a reverse-biased diode
   b. As a photodetector
   c. As an LED
   d. As a perfect insulator
   e. None of the above

6. What gives a double-heterostructure laser better efficiency than a homostructure laser?
   a. Reverse biasing
   b. Better confinement of stimulated emission in the active layer
   c. Restriction of current flow to the active layer
   d. Lower levels of spontaneous emission
   e. A homostructure laser is more efficient

7. Which type of semiconductor lasers emit a single longitudinal mode?
   a. Homojunction
   b. Buried-heterostructure
   c. Monolithic arrays
   d. Distributed-feedback
   e. Those built for compact disc players

8

8. If you want a distributed-feedback laser to emit light at 1000 nm, and the semiconductor has a refractive index of 3.2 at that wavelength, what should the spacing be for a first-order grating (m = 1)?
   a. 100 nm
   b. 143 nm
   c. 250 nm
   d. 500 nm
   e. 617 nm

9. Which family of semiconductor lasers have produced the highest power levels?
   a. GaAlAs
   b. InGaAs
   c. InGaAsP
   d. Lead-salt
   e. GaInP

10. Which family of semiconductor lasers has produced the shortest wavelength (in a device able to operate continuously at room temperature)?
    a. GaAlAs
    b. InGaAs
    c. InGaAsP
    d. Lead-salt
    e. GaInP

Other Lasers

ABOUT THIS CHAPTER

Not all lasers fit neatly into the three categories of gas, solid-state and semiconductor. In this chapter we will learn about three important types that are categories in themselves. One, the dye laser, is an important tool in scientific research and is finding growing uses in medicine. The other two, the free-electron laser and the X-ray laser, remain in development. At the end we will briefly mention a few other lasers.

TUNABLE DYE LASERS

Most lasers we have learned about emit light at fixed wavelengths. However, for many applications it would be helpful to be able to change or tune the laser wavelength. In some cases, the goal is to produce a precise wavelength not available from other lasers. In others, the goal is to scan a range of wavelengths. Some developers hope the vibronic solid-state lasers described in Chapter 7 will someday meet some of these needs. For now, however, the most broadly tunable laser is the organic dye laser.

The active media in dye lasers are organic dyes, a family of large and complex molecules. They are called dyes because they are brightly colored, a property that comes from their complex sets of electronic and vibrational energy levels. The vibrational energy levels create many sublevels of the electronic states, as in vibronic lasers. Thus, transitions between those energy levels can occur at an unusually broad range of wavelengths, allowing broadband laser oscillation. Different dye molecules can function as lasers over different ranges of wavelengths, as we will explain in more detail later.

Laser dyes normally are dissolved in liquid solvents, which serve as hosts for the active medium just as glasses or crystals are hosts for the active media in solid-state lasers. Dyes have produced laser light in both the vapor phase and when doped in transparent solids, but they don't work as well as in liquids. Dye lasers emit visible, near infrared, and near ultraviolet light, but no single dye has that broad a tuning range. To cover the entire visible spectrum, you must switch among several dyes, each with a limited tuning range.

Applications Influence Dye-Laser Design

With comparatively high gain, the dye laser is a versatile and flexible tool. Its performance can be adjusted in many ways, creating variations on the basic dye-laser theme. However, this versatility comes at the cost of complexity.

Like many other general-purpose tools, a dye laser is not a universal replacement for more specialized lasers. Dye lasers can generate continuous-wave beams at wavelengths not available from a helium-neon or argon laser, but if you can use a wavelength available from one of those lasers, you don't want a dye laser. You might think of a dye laser as an "all in one" screwdriver with a single handle and many blades, and helium-neon or argon lasers as standard screwdrivers. The all-in-one model is better for a variety of jobs that involve driving a few screws, but if you have to drive a lot of screws, or only one type, you do better with a standard screwdriver.

Because of its versatility, the dye laser has been adapted to meet specific needs. It is fair to say that the technology has been shaped by its applications. Before we examine dye lasers in more detail, let's look at their major uses.

Tunable Wavelength Output

The major application of dye lasers has long been to generate light that is tunable across the visible spectrum. Initially, virtually all dye lasers were used for spectroscopy, the measurement of the absorption and emission lines characteristic of various materials. Now a number of other applications have emerged that require light of specific wavelengths, especially in medicine, where specific wavelengths are needed for measurement and therapy. However, most dye lasers continue to be used in research and development.

All these applications require laser emission that can be adjusted in wavelength, usually across much of the visible spectrum, and often into the near-infrared or near-ultraviolet as well. This means that the laser cavity must include wavelength-changing optics.

The desired tuning ranges almost always span more than the range of a single dye, so the design must allow for switching of laser dyes.

Research lasers have some added requirements. They must allow adjustment of many characteristics, including power levels and spectral bandwidth, to match experimental requirements that can easily change. Many researchers are especially concerned with narrow bandwidth and stability. Some prefer continuous-wave emission; others prefer pulsed operation.

Medical lasers have different requirements. Lasers may be designed to perform specific procedures, such as treating birthmarks or shattering kidney stones. Physicians don't need the same flexibility as research scientists, but they do need instruments that are reliable and easy to use.

Short-Pulse Research

Another major use of dye lasers is in research on ultrashort light pulses. The attraction of dye lasers for this application is that their unusually broad gain bandwidth lets them generate exceptionally short pulses.

There is a fundamental relationship between the minimum possible duration of a laser pulse and the narrowest possible spectral bandwidth in that emission. The shorter the pulse, the broader must be the bandwidth. Conversely, the narrower the bandwidth, the longer the time the laser must operate to produce that bandwidth stably. The fundamental limit is given by:

$$\text{Pulse Length} = 0.441 \ / \ \text{Bandwidth}$$

where,
> pulse length is measured in seconds,
> bandwidth is measured in hertz.

Thus, if you want pulses that last only 10^{-9} second (one nanosecond), the laser's spectral bandwidth must be at least 441 megahertz. If you want picosecond (10^{-12} second) pulses, the laser bandwidth must be at least 441 gigahertz. This limitation applies to pulses generated by modelocked lasers, as well as to those from an unregulated laser.

This transform limit pushes researchers trying to generate the shortest possible pulses to use the laser with the broadest possible bandwidth. This laser is the dye laser, and some commercial models are designed expressly to produce such short pulses. Note also that it makes researchers who want extremely narrow spectral bandwidth use continuous-wave lasers.

Basic Structure of Dye Lasers

Figure 9-1 shows the internal structure of a generalized dye laser. One key element that may not be present in short-pulse lasers is the tuning optics. As we saw in Chapter 4, this is a prism or diffraction grating that deflects light of different wavelengths at different angles. It and the cavity mirrors are arranged so only one wavelength can oscillate in the laser cavity. Other wavelengths are deflected from the axis of the cavity and are not amplified, even if they fall within the laser's gain bandwidth.

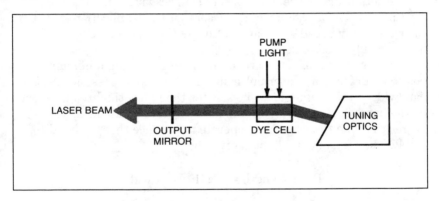

Figure 9-1. Structure of a generalized dye laser.

One key element always present is an optical pumping source. Laser dyes get their energy by absorbing light at wavelengths shorter than they emit. The bright light needed to produce a population inversion can come from various sources, including flashlamps and pulsed or continuous-wave lasers. As we will see later in this chapter, there are considerable differences among these light sources, so this illustration is very general. For continuous-wave dye lasers, which produce modest steady power levels, the pump light must be focused to a small volume within the dye.

The solution containing the laser dye is placed in the laser cavity. Our generalized picture shows a dye cell, without indicating if the cell is sealed or the liquid flows through it. In pulsed lasers with modest average power, the dye cell may be a sealed container called a "cuvette." At higher powers the dye solution may be pumped rapidly through the cell to prevent heating and degradation. In continuous-wave lasers, the dye may pass through the tightly focused pump-laser beam in an unconfined jet. (This avoids optical damage to the windows of the dye cell.) Laser dyes have high gain, so you do not need a long length of dye in the laser cavity.

Most dye lasers are designed with room for accessories inside and outside the cavity, including modelockers to reduce pulse duration, and optics to narrow the spectral linewidth. We don't show them here, but such accessories can play a vital role in operation of dye lasers.

In talking about dye lasers, we may make some functional distinctions that are not always apparent in commercial dye lasers. For example, the pump source in flashlamp-pumped dye lasers is built into the laser itself. Also, many dye lasers are sold with sets of accessories, and might be packaged as picosecond dye lasers or other specialized types. You should remember that these are packaging choices, not physical requirements.

Following our description of the functional elements of dye lasers, we will look in more detail at arrangements for tuning laser wavelength, the dyes themselves, and the various types of optical pumping.

Tuning and Cavity Design

Cavity design and tuning of dye lasers are closely tied. Except for some ultrashort-pulse lasers, cavities are designed to allow tuning. *Figure 9-2* shows three basic approaches to tuning dye-laser wavelength, with a diffraction grating, with a prism, or with some other wavelength-selective element such as a tunable etalon or a tunable filter. Actual laser pump cavities often have additional mirrors to direct pump and laser beams between components.

Prisms and diffraction gratings disperse light at different angles as a function of wavelength, so tuning with those elements requires turning either the wavelength-selective element or other optical components. In our examples (*Figures 9-2A and B*), the grating is

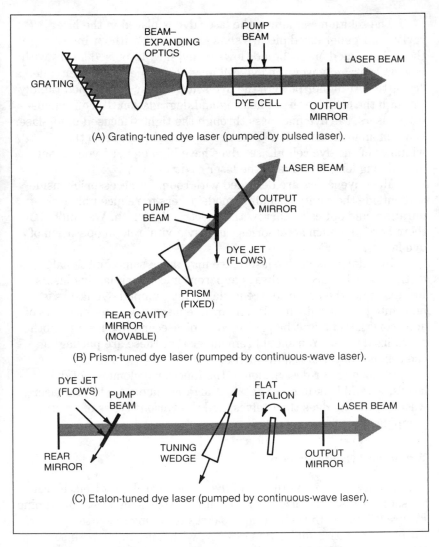

(A) Grating-tuned dye laser (pumped by pulsed laser).

(B) Prism-tuned dye laser (pumped by continuous-wave laser).

(C) Etalon-tuned dye laser (pumped by continuous-wave laser).

Figure 9-2. Three examples of wavelength-tunable laser cavities used with dye lasers.

turned to change the wavelength, but the prism stays fixed while the rear cavity mirror moves, to select a particular wavelength refracted by the prism.

In the third example in *Figure 9-2C*, the main tuning element is a wedge-shaped etalon. As we saw in Chapter 5, an etalon functions like a miniature optical cavity to limit oscillation to one range of wavelengths within the laser's gain bandwidth. In a tunable etalon, the distance between the two reflective surfaces is changed to tune the wavelength at which oscillation is possible. Normally the two surfaces are not moved back and forth, but the etalon is moved in some other way that changes the distance the beam travels between them. The wedge-shaped etalon can be moved up and down in the laser beam so the light travels through different thicknesses—and thus sees different spacings of the two reflective surfaces. An alternative is the fine-tuning flat etalon (at right), which is rotated to change the distance the laser beam travels between the surfaces.

Another approach to tuning, not shown here, is the use of a birefringent filter. It is placed in the laser cavity, and turned to change the wavelength that can oscillate.

Most laser dyes have gain bandwidths from 10 to 70 nm. Even without a tuning element, the naturing line-narrowing inherent in laser oscillation will limit linewidth to a few nanometers. Simple tuning prisms and gratings limit the range of wavelengths to a much smaller range, several thousandths of a nanometer. The addition of an etalon can limit oscillation to an even narrower range of wavelengths. With active control systems and sophisticated additional optics, the linewidth of a continuous-wave dye laser can be reduced well below one megahertz—less than one part in 10^9 at visible wavelengths, or under 5×10^{-7} nm.

The designs of dye-laser cavities are heavily influenced by the need to tune their output wavelength. This takes dye lasers beyond the standard straight-line cavity, where two mirrors are on opposite ends of a laser rod or tube. We simplified the cavities in *Figure 9-2* to make it clear how they worked, but most dye-laser cavities include extra mirrors for focusing and beam direction because the cavity components are not in a straight line. Even more complex arrangements are sometimes used to optimize performance, such as the ring cavity shown in *Figure 9-3*. At first glance, you might think light could pass in either direction around the ring, but this design contains components that prevent oscillation in one direction. Again, this drawing has been simplified by removing some components used in commercial lasers. The ring cavity limits laser oscillation to a narrower range of wavelengths than other cavities

As we indicated earlier, organic dyes have high gain under

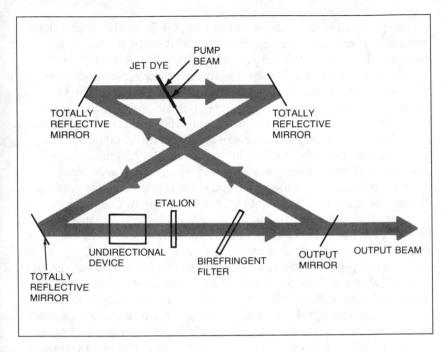

Figure 9-3. A ring dye laser.

proper pumping conditions, so the cavity contains only a short length of active medium. The high gain of dyes makes them suitable for use in oscillator-amplifier configurations when high power is required.

We should note that the tuning techniques and cavity designs we have discussed here can be used with any type of tunable laser. In practice, they are used with dye lasers almost exclusively because dye lasers are the most tunable lasers.

Properties of Organic Dyes

The organic dyes used in lasers are large and complex molecules with multiple ring structures and complex spectra. The dye molecules are fluorescent, meaning that they can absorb a short-wavelength photon and almost immediately "fluoresce," or emit light at a longer wavelength. This fluorescence occurs without a population inversion. Light from an external source excites the molecule, which then drops back down the energy-level ladder, emitting light to release energy.

This is the process by which minerals and other materials fluoresce—emitting visible light when illuminated by shorter-wavelength ultraviolet light.

In a dye laser, the molecules are excited by a light bright enough to produce a population inversion on the laser transition. The upper laser levels typically have lifetimes of a few nanoseconds, giving them high gain but low energy storage. The laser emission is on electronic transitions which have vibrational sublevels, as shown in *Figure 9-4*. We have seen similar effects in other lasers, like the carbon-dioxide gas lasers, where rotational sublevels of vibrational transitions create many distinct lines. In organic dyes, the levels are so closely spaced that they merge together to form a continuum. Instead of tuning a dye laser's wavelength in steps, you can adjust it continuously.

Light excites a dye laser, and much of the absorbed energy

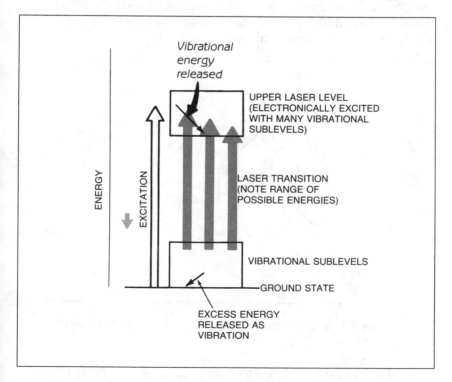

Figure 9-4. Energy levels in a typical laser dye.

ultimately emerges as light. After dye molecules are excited from their ground state, they drop down to the lowest vibrational sublevel of the electronically excited state. The laser transition takes the molecule to a vibrationally excited level of the electronic ground state. The remaining energy is dissipated as heat. As you can see in *Figure 9-4*, the laser transition can be to various levels within a range defined by the vibrationally excited sublevels of the ground state, giving it broad gain bandwidth.

The wavelengths emitted by dyes depend on their chemical composition. Most dyes belong to several families of organic compounds, and seemingly countless variations are possible just by adding an atom or two to different places on the molecule. Developing laser dyes is a serious exercise in organic chemistry, and you don't want to worry about the chemical names or the details. *Figure 9-5* gives you an idea of what you want to miss—it's the basic structure of the Coumarin family of laser dyes. The question marks in *Figure 9-5* indicate where different atoms can be added to make new molecules. There are about 100 coumarin dyes.

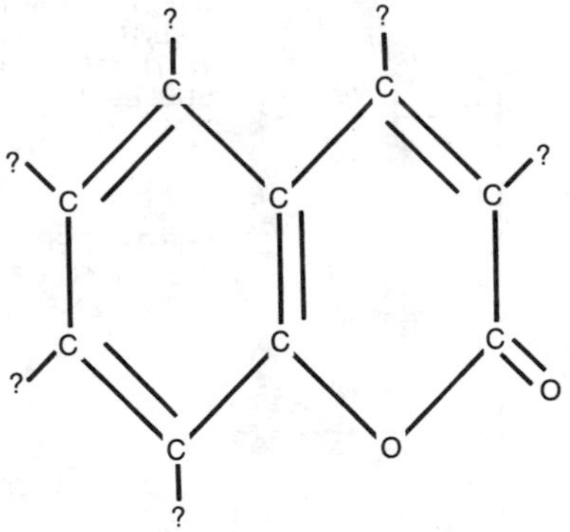

Figure 9-5. Basic chemical structure of Coumarin family of laser dyes.

Most laser dyes have gain bandwidths of 10 to 70 nanometers, with typical ranges 20 to 40 nm. Gain is the highest at the center of the range, and drops at longer and shorter wavelengths. Although individual dyes fall far short of covering the entire visible range, groups of dyes can combine to provide continuous coverage. Once you move to the edge of the tuning range of one dye, you switch to another dye. In practice, single dyes have enough range for many purposes, but some lasers are made with groups of dye cells that can be switched into position successively one after another to tune beyond the range of any single dye.

The range of wavelengths a dye laser can emit depends on the wavelength, power, and pulse characteristics of the pump source. Often the differences are small, but certain pump sources can work much better than others for specific dyes. For example, many dyes will not lase when driven by a continuous-wave laser, but can lase when pumped by a pulsed light source.

Organic dye molecules are more fragile than other laser media, and most degrade after tens or hundreds of hours of use. The bright light needed to excite them to the upper laser level also can break up the molecules. This problem also makes it important to keep dye solutions cool, especially if the laser is operated at high power. Like other chemical reactions, dye decomposition occurs faster at warmer temperatures. For this reason, all but the lowest-power dye lasers normally have systems that pump the dye solution through the light-emitting zone. In continuous-wave lasers, the dye flows through in a continuous jet.

Like other complex organic molecules, most dyes do not dissolve readily in water. Dye solutions normally use organic solvents like methanol, ethanol, and other liquids that may be both toxic and flammable. The dyes themselves can be highly toxic.

Flashlamp-Pumped Dye Lasers

Many laser dyes can be pumped by broadband "white" light from flashlamps. There are two basic types of flashlamp-pumped dye lasers. In *Figure 9-6A* a linear flashlamp transfers its energy to a parallel linear dye cell, much like a flashlamp pumps a solid-state laser rod. In *Figure 9-6B* the dye flows through the center of a coaxial flashlamp, which transfers its energy directly to the dye inside. For simplicity, the dye flow system is not shown here.

The length of pulses from a flashlamp-pumped dye laser depends on the duration of the flashlamp pulses, which are typically

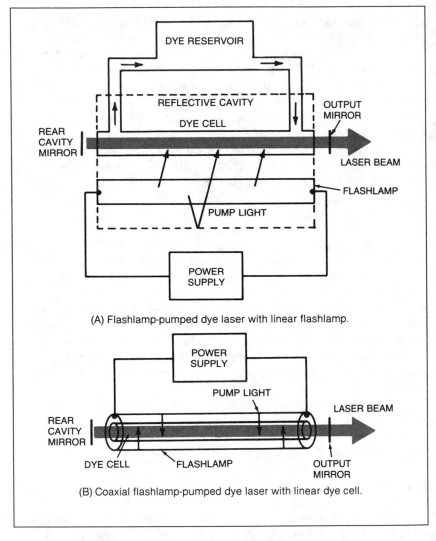

(A) Flashlamp-pumped dye laser with linear flashlamp.

(B) Coaxial flashlamp-pumped dye laser with linear dye cell.

Figure 9-6. Linear and coaxial flashlamp-pumped dye lasers.

about a microsecond, but last up to 500 microseconds. Flashlamp-pumped dye lasers can be designed either to produce a few powerful pulses a minute, a few hundred low-energy pulses per second, or some intermediate value.

The main advantages of the flashlamp-pumped dye laser are its ability to produce pulses with higher energy and higher average power than other dye lasers and its modest capital cost. It is used in the research laboratory and for some specialized medical treatment.

Pulsed Laser Pumping

Pulsed lasers also can pump dye lasers. *Figure 9-7* shows how a single pulsed laser can pump both a dye oscillator and an amplifier. Focusing optics spread light from the pump laser out to illuminate the entire side of the dye cell. In our illustration, the dye beam is produced at right angles to the pump beam. A right angle is not necessary, but pulsed lasers normally do not pump along the length of the dye laser cavity.

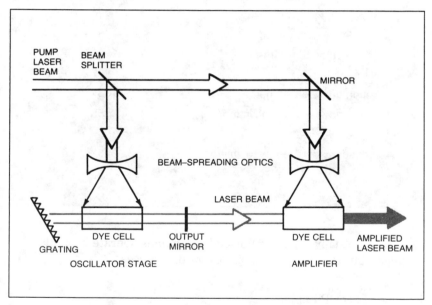

Figure 9-7. Pulsed laser pumping of a dye oscillator-amplifier.

In the oscillator-amplifier configuration we show, the pump beam is split at a beamsplitter. Part of the beam pumps the dye cell in the oscillator, and the other part pumps the dye cell in the amplifier stage. This arrangement can provide higher power than an oscillator alone.

Although many pulsed lasers are available, the choices for pumping dyes are limited by a fact we mentioned before: the pump wavelength must be shorter than the dye emission wavelength. Because most dye lasers are operated at visible wavelengths, this limits pump wavelengths to the ultraviolet and the short-wavelength end of the visible spectrum. The major wavelengths are given in *Table 9-1.*

Table 9-1. Wavelengths for Pulsed Laser Pumping

Pump Laser	Wavelength (nm)
Krypton-fluoride excimer	249
Frequency-quadrupled neodymium	266
Xenon-chloride excimer	308
Nitrogen	337
Xenon-fluoride excimer	351
Frequency-tripled neodymium	355
Copper vapor	510 & 578
Frequency-doubled neodymium	532

Conceptually, all these pulsed laser pumps operate in similar manners, but there are some important differences.

Excimer Laser Pumping

Excimer laser pumps can generate high-power ultraviolet pulses at reasonable repetition rates in the ultraviolet. Their wavelengths are short enough to pump dyes throughout the visible spectrum, but the shorter excimer wavelengths tend to shorten dye lifetimes. Capital costs are higher than for flashlamp-pumped lasers, but excimer lasers are more versatile and easier to operate.

Neodymium Laser Pumping

The 1064-nm fundamental wavelength of neodymium-YAG is too long to pump any dyes except a handful that generate longer infrared wavelengths. However, the second, third, and fourth harmonics of neodymium are short enough to pump visible-wavelength dyes. The 532-nm green second harmonic can pump dyes at wavelengths as short as 539 nm, but the third harmonic can cover the rest of the visible and part of the ultraviolet. The fourth harmonic is used only to pump a few short-wavelength ultraviolet dyes.

The functional characteristics of pulsed neodymium-laser pumps and excimer pump lasers are similar. Both have similar repetition rates and output powers. Both have higher capital costs and lower operating costs than flashlamp-pumped dye lasers, and both can be used for other purposes than pumping dyes.

Nitrogen Laser Pumping

The 337-nm nitrogen laser was the first pulsed ultraviolet laser used to pump dye lasers. However, its pulse energy is limited, and it fell out of favor after the introduction of more powerful excimer and pulsed neodymium laser pumps.

Small nitrogen lasers can be made inexpensively, and this has led to the introduction of pulsed dye lasers pumped with small nitrogen lasers. A complete package with pump laser and dye costs only a few thousand dollars, making it attractive for use in instruments or in laboratories with very limited budgets.

Copper Vapor Laser Pumping

The repetitively pulsed green and yellow lines of the copper vapor laser can effectively pump dyes at wavelengths longer than about 530 nm. The rapid sequence of pulses can generate high average power less expensively than an argon laser can generate high continuous-wave powers, but the repetitive pulses are not suited for all applications.

Continuous-Wave Laser Pumping

Dye lasers can be pumped by continuous-wave argon or krypton lasers to generate a continuous-wave beam. Designs of continuous-wave dye lasers differ markedly from those of pulsed dye lasers.

Because the available pump power is modest, the pump beam is tightly focused, generating a high power level in a very small area. The dye is in a thin flowing jet, which is not housed in a glass tube to avoid optical damage to the tube from the concentrated pump laser power. The rapid flow of the dye solution also keeps it from being heated to excessive temperatures.

We showed two simple designs for continuous-wave dye lasers in *Figure 9-2*. We should warn you that those are simple designs, without the "bells and whistles" needed for extremely narrow linewidth in high-performance models. Even the more elaborate ring laser design in *Figure 9-3* is a simplified version of the types used in commercial lasers.

Some measurement techniques require continuous-wave beams. Another important attraction of continuous-wave dye lasers is the potential to emit a very narrow range of frequencies for ultraprecise measurements. We saw earlier that the minimum bandwidth possible in a beam of light depends on the length of time the light is emitted:

$$\text{Bandwidth} = 0.441 \text{ / Pulse Length}$$

where,
> bandwidth is in hertz,
> pulse length is in seconds.

Thus, the longer the laser emits light, the narrower the minimum bandwidth. Continuous-wave dye lasers do not allow infinitely narrow linewidth, but commercial dye lasers have gone below 1 MHz (about one part per billion in frequency), and even lower figures have been recorded in the laboratory.

Ultrashort Pulses

If we turn our formula around and remember that laser dyes have an especially broad gain bandwidth, we can see another possible use of the dye laser—to generate ultrashort pulses. In this way, we can see that the minimum possible pulse length is given by

$$\text{Pulse Length} = 0.441/\text{Bandwidth}$$

To generate the shortest possible pulses, we need to use as much as possible of the dye's bandwidth, with no tuning optics in the laser cavity. You have to convert bandwidths measured in nanometers into the equivalent in hertz in three steps. First find the relative bandwidth, by dividing the emission bandwidth in nanometers ($\Delta\lambda$) by laser wavelength (λ)

$$\text{Relative Bandwidth} = \Delta\lambda / \lambda$$

Then you convert wavelength in nanometers into frequency in hertz (ν), using the earlier formula

$$\nu = c / \lambda$$

where,
 c is the speed of light (300,000 km/sec).

Then you multiply the frequency by the relative bandwidth (which is a fraction) to find the bandwidth in hertz. Step through this for a dye laser with 1-nm emission bandwidth at center wavelength of 500 nm, and you find the minimum theoretical pulse length is 367.5 femtoseconds (367.5×10^{-15} second). You can make a dye laser's emission bandwidth somewhat broader, but you can't directly produce laser emission across the full gain bandwidth.

The usual way to generate short pulses from a dye laser is by a technique called "synchronous modelocking." The dye laser is modelocked and pumped with a modelocked neodymium or argon laser, so it generates modelocked pulses synchronized with those of the pump laser.

Special techniques can squeeze pulses from a dye laser down to even shorter durations if they spread the light over a broader range of wavelengths than in the dye laser's output. The approach which has produced the shortest pulses is to pass very short pulses from a dye laser through a short length of optical fiber, which spreads them over a wider range of wavelengths. Then the pulses are reflected back and forth between gratings or prisms to squeeze them down in time. This has yielded pulses as short as 6 femtoseconds (6×10^{-15} second).

Harmonic Generation

Nonlinear crystals can generate harmonics of dye laser light, just as they do for other lasers. Because the dye laser wavelength is tunable, so are the harmonics. In practice, it is the second harmonic that is generated, and that generally lies in the ultraviolet. Because the nonlinear effects that generate harmonics are stronger at higher power levels, harmonic generation normally is done with pulsed dye lasers.

FREE-ELECTRON LASERS

The free-electron laser differs in many ways from the other types of lasers we have learned about. The active medium is a beam of "free" electrons—unattached to any atom—that is passing through a special type of magnetic field that varies periodically in space. The easiest way to view the magnetic field is by assuming it was created by an array of permanent magnets arranged with alternating polarity as shown in *Figure 9-8*. If the north pole of the first magnet is above the electron beam, as shown in the illustration, the south pole of the second magnet must be above the electron beam, then the north pole of the third, and so on.

In *Figure 9-8*, the electron beam enters the magnet array at a slight angle. The magnetic field from the first magnet bends the electron beam in one direction, then the opposite-polarity field from the second magnet bends it back in the other direction. This process repeats until the electron beam passes out the other end of the magnet array, which is called a "wiggler" or "undulator" because of its effect on the electron beam.

Electrons wiggling through the wiggler magnet field release some of their energy as light. Much of that light energy is, in essence, reabsorbed when the electrons wiggle back the other way. However, light can be amplified if its wavelength (λ) meets an approximate resonance criterion:

$$\lambda = p \, / \, 2[1-(v^2 \, / \, c^2)]$$

where,
  p is the length of a wiggler-magnet "period" (north-to-south and back),
  v is the electron velocity,
  c is the speed of light.

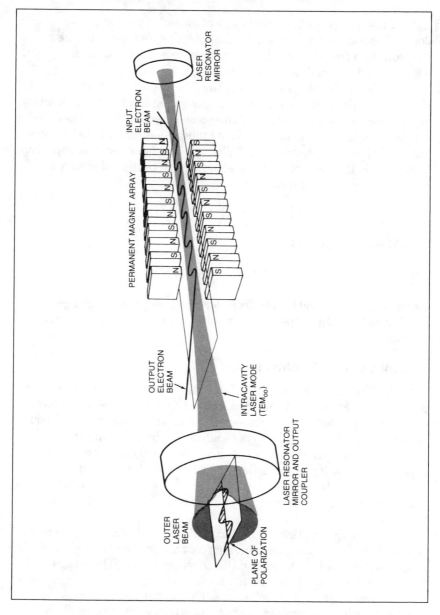

Figure 9-8. Structure of a free electron laser. *(Courtesy Luis Elias, University of California at Santa Barbara Quantum Institute)*

If that $[1-(v^2/c^2)]$ reminds you of formulas you've seen describing the theory of special relativity, there's a reason. The electrons in a free-electron laser are moving so fast that their behavior is described in relativistic terms.

While the previous version of the formula is helpful in highlighting the basic physics of a free-electron laser, a more practical version incorporates the ratio of the accelerated electron's mass and acceleration energy (E, measured in million electronvolts) to its rest mass, 0.511 million electronvolts. The acceleration energy is important because it is the normal way of measuring the output of electron-beam generators. The formula is

$$\lambda = p \,/\, 2[(0.511 + E)\,/\,0.511]^2$$

or, to simplify things a bit,

$$\lambda = 0.131\, p \,/\, (0.511 + E)^2$$

As the formula indicates, the free-electron laser wavelength gets shorter as the magnet period decreases and as the electron energy increases.

Tunability of Free-Electron Lasers

We have already said that electrons in a free-electron laser are not bound to atoms, and thus don't have fixed energy levels. Look carefully at our formula for free-electron laser wavelength, and you can see that there are two ways to adjust it—by changing the wiggler-magnet period, or by changing the electron energy (which depends on velocity). In practice, it's usually easier to change the electron energy than the structure of the wiggler field, but the important fact is that free-electron lasers are inherently tunable in wavelength.

Even more exciting is the fact that the wavelength range is, in principle, extremely broad, ranging from the microwave region at the long-wave end to soft X rays on the short-wave end. This might make it seem that you need only turn a knob to adjust electron energy and get the whole range of wavelengths. That is not the case, because practical design constraints limit any one free-electron laser to a much more limited range of wavelengths. Nonetheless, that range is broad compared with other tunable lasers.

Types of Wigglers

Figure 9-8 shows the wiggler as a stack of magnets of alternating polarity. This is the simplest way to view a wiggler, but it can be somewhat misleading. The key element of a wiggler is not the magnet per se, but the magnetic field it generates. It is the magnetic field which bends the paths of the electrons passing through it. The field direction passes through a complete cycle as you move from north-to-south to south-to-north magnets, then back to a north-to-south magnet. So you can also view the magnetic field as a sinusoidal wave, which oscillates as you move along the wiggler.

This has led to another approach to wiggler design—using a strong electromagnetic wave to provide the sinusoidally varying magnetic field. While this approach may not seem as straightforward as using a magnet array, it has its own advantages. Recall that free-electron laser wavelength decreases with period of the wiggler magnet. With physical magnets, it gets harder to build the array as the period becomes smaller. By using electromagnetic waves, you can produce very small periods, letting you generate shorter free-electron laser wavelengths without the need for more energetic electrons (which, as we will soon see, require more powerful accelerators).

The use of electromagnetic-field wigglers also opens up some intriguing possibilities in bootstrapping free-electron lasers to shorter wavelengths. Present technology can generate more power from free-electron lasers at longer wavelengths than at shorter wavelengths. Suppose you start with a permanent-magnet wiggler that generates a powerful beam at 100 micrometers in the far infrared, then use the magnetic field of that laser beam as the "wiggler field" to generate a shorter-wavelength laser beam. We show this idea in *Figure 9-9*. With a comparatively modest electron energy of 10 million electronvolts, you could generate a second-stage free-electron laser wavelength of 119 nanometers in the ultraviolet—nearly a factor of 1000 shorter than the first-stage wavelength. Two-stage free-electron lasers are not that simple in practice, but they nonetheless offer intriguing possibilities for generating short-wavelength, tunable laser beams.

In drawing our simple picture of the free-electron laser, we assumed that the magnetic field had uniform amplitude along the entire wiggler. This case is the simplest to analyze, but is not necessarily the best. The efficiency with which free-electron lasers

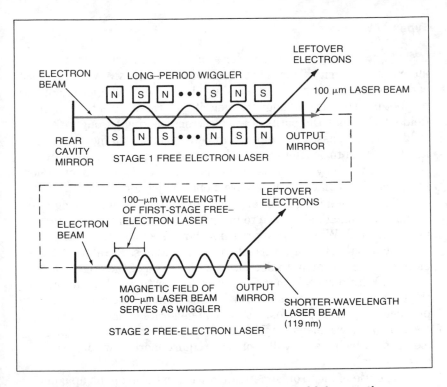

Figure 9-9. A two-stage free-electron laser which uses the magnetic field from a 100-μm free-electron laser as the wiggler field for a second stage that generates a 119-nm beam. (We've bent the 100-μm beam for artistic convenience—the lasers wouldn't fit on the page in a straight line.)

extract energy from the electron beam is increased if the period gradually changes along the wiggler. The reason is that the electron beam loses energy as it passes through the wiggler, so it can slip out of resonance with a fixed-period field. Changing the period by the proper amount helps the electron beam and wiggler stay in resonance through the entire wiggler, so more laser energy can be extracted from the electrons.

Electron Accelerators

One strength of free-electron lasers is that they draw upon a well-developed technology outside of the normal realm of laser physics: charged particle accelerators. The technology had its roots in early atom smashers, and still plays a vital role in particle physics research. The Department of Energy plans to build the world's largest particle accelerator for research in fundamental physics, the multibillion-dollar Superconducting Supercollider. Particle accelerators also are used in some areas outside of research labs.

There are many types of charged particle accelerators, but they all share the same basic principle. They pass an electrically charged particle through an electric potential that accelerates it to higher velocity, thus giving it more energy. For example, a negatively charged electron is attracted toward a positive potential. Accelerators also work with protons, ionized atoms, or charged subatomic particles such as muons. The particle energy normally is measured in electronvolts, where one electronvolt is the amount of energy that an electron acquires when falling through a potential of one volt. The electron accelerators used for free-electron lasers generate particles with energies of millions of electronvolts.

The electron beams generated by accelerators are pulsed, with the length of the pulses ranging from hundreds of microseconds to picoseconds. The longer pulses typically contain much shorter subpulses. Pulses may be single or repetitive. Pulse characteristics can strongly affect lasing characteristics. So can other characteristics of the electron beam, including the parallelism of the paths of the electrons, and their spread in energies. As with a laser beam, the quality of an electron beam degrades with an increasing spread in energy and direction.

We won't talk about the details of accelerator design, but we should briefly describe three major types used with free-electron lasers. (Many others are used for other purposes.) The first type used in a free-electron laser was the linear accelerator, in which all the elements lie in a straight line, and the electron travels a long, linear path through the accelerator. After the electron beam passes through the undulator, it can be directed to an energy-recovery system or just dumped to a beam dump, as shown in *Figure 9-10A*.

One alternative is the storage ring, in which electrons accelerated to a high velocity travel around a closed loop, confined by strong magnets. If one arm of this loop contains an undulator, as shown in *Figure 9-10B*, it can drive a free-electron laser. Because the electrons pass repeatedly through the wiggler, this promises more efficient operation. Unfortunately, passage through the wiggler

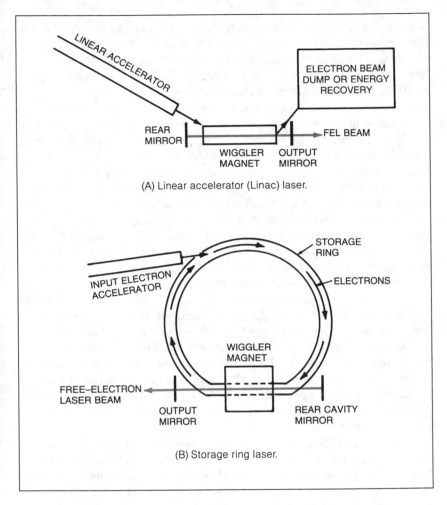

(A) Linear accelerator (Linac) laser.

(B) Storage ring laser.

Figure 9-10. Design of free-electron lasers driven by accelerators.

removes some electron energy, degrading beam quality and taking the electrons out of resonance with the wiggler field. Extra beam handling equipment is needed to take care of this problem.

Another alternative is the electrostatic or Van de Graaff generator, which accumulates a large static charge, then discharges it as a pulse of electrons which pass through a wiggler. This approach is attractive because it recovers most energy that the electrons retain after passing through the magnet. It works best for long-wavelength free-electron lasers.

Operating Regimes

Free-electron lasers operate over such a broad range that the internal interactions can be quite different. In practice, these are divided into two operating regimes. One is the "Compton" regime, best described as involving interactions between individual particles. The other is the "Raman" regime, best described by collective, multiparticle effects. The difference is important because the operating characteristics can differ dramatically.

Free-electron lasers operate in the Compton regime if the electron energy is high (20 million electronvolts or more), the current flow low, and the wavelengths shorter than about 20 micrometers at the short end of the infrared spectrum. In this regime, there is an optimum wiggler length for maximum gain; gain will drop if the wiggler is too long or too short. Gain typically is low, and present only in one direction, when light waves and electrons are travelling in the same direction. There is no gain when the light passes through the cavity in the opposite direction.

Free-electron lasers that operate in the collective regime have lower electron energy (typically under 5 million electron volts) and higher current density. They also operate at wavelengths longer than about 100 micrometers. Gain is higher in the collective regime, and can be exponential with distance travelled through the wiggler. Thus, output powers have been higher.

Cavity Designs

As we saw earlier, ordinary low- and high-gain lasers use different cavity designs. The same is true for free-electron lasers operating in the Compton and Raman regimes.

The low gain in the Compton regime makes low cavity loss essential. This requires a rear-cavity mirror with high reflectivity and an output mirror that lets only a small fraction of the light out of the cavity. The technology for making high-reflectivity visible mirrors is well-developed, but those mirrors don't stand up well under high powers, or when exposed to the intense ultraviolet light also generated in a visible-wavelength free-electron laser.

The high gain of the Raman regime makes cavity losses a minimal concern, so the output mirror can transmit much of the light from inside the laser cavity. This makes the optics problem much more manageable, and permits generation of very high-power output pulses. Unfortunately, the far-infrared wavelengths generated by Raman-regime free-electron lasers have limited applications.

Large and Small Free-Electron Lasers

The biggest single attraction of the free-electron laser is its potential to generate truly awesome powers with high efficiency. Developers envision overall efficiencies of 20% to 50% or more, depending on how much energy can be recovered from the electron beam after it has passed through the free-electron laser. Such efficiencies are extremely high by laser standards.

When these high efficiencies are combined with the high powers available in electron beams, they open prospects for extremely high-power lasers. The technology for generating energetic particle beams is well developed, although currents usually fall short of those required for extremely high-power free-electron lasers. Peak pulse powers of a gigawatt have been generated at a one centimeter wavelength by a Raman regime free-electron laser. Prospects look reasonable for generating average laser powers in the megawatt range.

The main potential use for ultrahigh-power free-electron lasers is as weapons. The Strategic Defense Initiative hopes to demonstrate a massive high-power free-electron laser for potential use in defense against ballistic missiles. The laser, based on a mountaintop, would send its beam to a relay satellite, which would pass the beam on to other relay mirrors that would eventually deliver it to its target.

Meanwhile, other developers are working on small free-electron lasers. They include John M. J. Madey, the physicist who built the first free-electron laser at Stanford University. Their goal is to build compact lasers that could be tuned between about 1 to 10 micrometers, for medical research and treatment. Few lasers are now available between those wavelengths, and the developers hope that free-electron lasers could open new possibilities in medical treatment. This work, too, is supported by the Strategic Defense Initiative—but only because Madey and others lobbied Congress to set aside money for it.

Promises and Problems for Free-Electron Lasers

Free-electron lasers have exciting promise in many areas. They have impressive potential for efficient, high-power output. Their tunability can open new parts of the spectrum to laser research, especially short wavelengths where laser sources have been scarce or unavailable. Their combination of tunability and power makes them promising for many applications in medicine and other areas that may require reasonable power levels at specific wavelengths not available from other sources.

However, any realistic assessment must be tempered by looking at potential problems. So far, the highest powers have been demonstrated in the long-wavelength regime, not at more-useful shorter wavelengths. There are big problems with the short-wavelength optics needed to make a good laser resonator.

Perhaps most importantly, the technology of free-electron lasers remains young. Although Madey's first free-electron laser operated over a decade ago, free-electron laser experiments are large in scale and thus particularly time-consuming. Accelerators are costly, so for several years researchers had to use existing accelerators not designed for free-electron lasers. It will take more time to build special equipment and test how well free-electron lasers scale both to higher and lower powers. That step is critical to the success of the technology, but its outcome is never assured.

X-RAY LASERS

Laser researchers have long tried to push to ever-shorter wavelengths. Townes started out with the microwave maser, then the laser jumped to visible wavelengths—roughly a factor of 10,000 shorter. For two decades after the first ruby laser, researchers seeking shorter wavelengths made only limited progress into the ultraviolet. Only in the past decade has real progress been made on lasers at extremely short wavelengths in the "soft" X-ray part of the spectrum.

X-ray lasers differ in many ways from the other types we've described so far. Differences between X rays and longer-wavelength electromagnetic radiation lead to important differences in structure and operation. Some important X-ray laser research is highly classified, because it depends on small nuclear explosions. We will briefly explore this area (including some speculations on what's hiding behind the TOP SECRET stamps) after we take an initial look at X-ray physics.

X-Ray Physics

Earlier, we learned that visible and ultraviolet lasers operate on electronic transitions. X rays also are produced by electronic transitions, but of a different type. The transitions that produce visible or ultraviolet light involve the outermost electrons attached to an atom or molecule. Those electrons experience only a modest attraction from the atomic nucleus. For light elements such as hydrogen and helium, where only the outer electrons are present, the nuclear charge is low. In heavier elements such as iron or argon, the outer electrons are shielded from the much larger nuclear charges by complete shells of inner electrons. As a result, outer-shell transition energies are usually in the 1- to 10-electronvolt range. (We use the electronvolt energy scale because it is a handy way to measure energy in the X-ray range. To convert to wavelength (λ), you can use the handy formula:

$$\lambda = 1.24 \times 10^{-6} / \text{Energy}$$

where,

λ is in meters,
Energy is in electronvolts.

A 1-eV transition is at a wavelength of 1.24 micrometers; a 10-eV transition is at 124 nanometers.)

Inner-shell electrons are much closer to the positively charged nucleus, and without as much shielding they experience much more pull from the nuclear charge, as shown in *Figure 9-11* for selenium, which has 34 protons in its nucleus. That means that it takes a lot of energy—100 eV or more—to pluck an electron out of an inner shell and raise it to an outer energy level. Conversely, if an electron drops from an outer shell into a vacancy in an inner shell, as shown in *Figure 9-11*, it will release a highly energetic photon. The electromagnetic radiation emitted and absorbed by such transitions is called X rays.

We should warn you that there is no rigid boundary between X

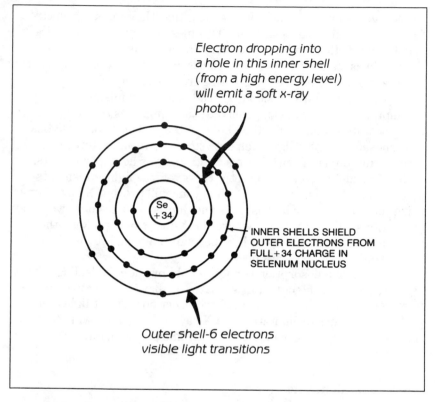

Electron dropping into
a hole in this inner shell
(from a high energy level)
will emit a soft x-ray
photon

Se
+34

INNER SHELLS SHIELD
OUTER ELECTRONS FROM
FULL+34 CHARGE IN
SELENIUM NUCLEUS

Outer shell-6 electrons
visible light transitions

Figure 9-11. X-ray transitions in selenium involve inner-shell electrons.

rays and ultraviolet rays. The two parts of the spectrum blend together, and definitions often vary. Some specialists call wavelengths longer than 10 nm "soft X rays," but others prefer to call them the "extreme ultraviolet." (In my more cynical moments, I suspect the terminology is chosen to fit the individual's desire to claim an "X-ray laser," or to avoid security restrictions on publishing "X-ray laser" papers.) The important point to remember is that there is a gradation of energy levels, some clearly ultraviolet, some clearly X ray, and some on the hazy middle ground.

Because they carry much more energy than photons of visible light, X rays have much stronger impact on what they strike. From a biomedical standpoint, X rays (and short ultraviolet wavelengths) are considered "ionizing radiation," meaning that they can strip electrons from atoms or molecules they hit. This means they can have a direct and dramatic effect on biomolecules. X rays are used in cancer therapy because they can damage cancer cells. However, they also can increase the risk of cancer originating from healthy cells.

Superman's X-ray vision may lead you to the mistaken assumption that X rays can penetrate anything. It is true that X rays can penetrate tissue better than bone or teeth. This is because tissue is composed mostly of light elements such as hydrogen, oxygen, and carbon, that do not absorb X rays as strongly as heavier elements, such as calcium in bones and teeth. However, the lighter elements absorb some X rays, and even ordinary air will prevent X rays from going too far. (That's why hospitals and dentist's offices observe X-ray precautions only in the room where the X-ray is being taken—the X rays don't penetrate further.) If Superman really had X-ray vision, his world would be a very dark one.

This strong absorption is not the only problem in dealing with X rays. At X-ray wavelengths, most materials have a refractive index very close to one, so they can neither reflect nor refract light well, putting some serious limitations on X-ray optics. As we will see later, there has been progress in dealing with this problem, but it is not fully solved.

X-Ray Laser Concepts

As with other types of lasers, an X-ray laser requires a population inversion. However, an X-ray population inversion is very hard to produce. Potential upper laser levels have extremely short lifetimes, and drop back to the lower level very quickly. Moreover, it takes a lot of pump energy to raise each atom to the excited level, so it takes a tremendous energy (or peak power) to produce a population inversion. This narrows the range of options for pumping an X-ray laser to essentially one: brute force.

The brute force approach is the epitomy of what one former X-ray laser researcher called the "telephone pole" theory of lasing. In the early days of laser research, he recalled, some people believed you could make anything lase—even a telephone pole—if you hit it hard enough. It took a long time before anyone learned how to apply enough brute force to make an X-ray laser.

To produce a population inversion at longer wavelengths, you can get away with raising one electron in each atom to the right excited state. However, to produce an X-ray population inversion you must do a lot more. You must strip many electrons from atoms, then wait (but not very long) for the atoms to recapture the electrons. The population inversion occurs as the electrons are being recaptured, before they drop all the way down to the inner electron shells.

Depositing this much energy vaporizes the X-ray laser material. However, for a brief instant before it dissipates, it forms a hot plasma that can emit an X-ray laser pulse. Researchers so far have found two variations on this approach, differing in the nature and scale of energy deposition.

Bomb-Driven X-Ray Lasers

It's hard to imagine more brute force than that provided by a nuclear explosion. That's what researchers from the Lawrence Livermore National Laboratory in California used in a series of supersecret and highly controversial X-ray laser experiments at the Department of Energy's Nevada Nuclear Test Site. The intense burst of X rays from the nuclear fireball excited the X-ray laser medium, generating intense X-ray laser pulses. Some observers have claimed that the process isn't really a nuclear "explosion," but that seems a largely academic distinction. For the nuclear reaction to generate an intense burst of X rays, it must be producing enough energy to heat the material that emits the X rays very hot.

That X-ray laser design is being developed as part of the Strategic Defense Initiative, or "Star Wars" program. Its attraction is the ability to concentrate energy from the nuclear blast and direct it long distances, instead of letting it spread out into space unimpeded, as it would from a conventional bomb. Developers called the concept a "third generation nuclear weapon." However, details remain classified. The program also is at the center of intense controversy over the potential effectiveness of such weapons, the desirability of using nuclear weapons (even if only "small" ones) to defend against nuclear weapons, and the accuracy of interpreting still-secret research results.

Laser-Driven X-Ray Lasers

A parallel unclassified program conducted by Livermore uses the different pumping approach shown in *Figure 9-12*. Researchers zap a thin metal foil with a short, high-energy pulse from a massive conventional laser built for fusion research. The fusion laser pulse contains trillions of watts but only lasts about a nanosecond (10^{-9} second). The intense laser light strips electrons from atoms in a linear segment of the metal foil, creating a linear plasma. As the electrons recombine with the atoms, they emit a pulse of coherent X rays.

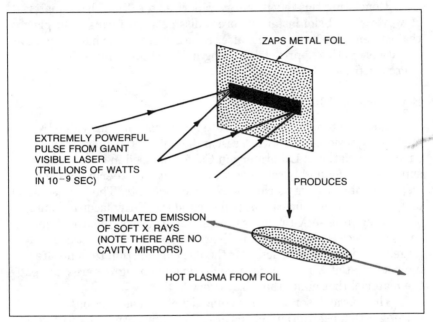

Figure 9-12. Stimulated emission of X rays from a laser-produced plasma.

Enough results have been published to indicate that this approach does generate X rays by a process that at least looks like stimulated emission. However, output powers remain modest compared to the amount of pump power needed to generate the X rays.

The X-Ray Mirror Problem

As we mentioned earlier, X-ray mirrors are a problem. Fortunately, gain is high enough that they aren't essential for X-ray laser emission. However, some specialists would quibble with our definitions, pointing out that the X-ray "lasers" described so far are not resonant-cavity oscillators, but only provide X-ray amplification by stimulated emission, and thus shouldn't be called lasers.

More importantly, usable X-ray resonator mirrors might help improve directionality and output power of the X-ray beam, particularly from the laser-pumped unclassified devices. Progress is being made by making multilayer X-ray reflectors, similar to the multilayer reflective coatings used for visible-light optics. They are made by depositing extremely thin alternating layers of two materials with slightly different X-ray refractive indexes. Although reflection at each boundary is small, the sum over many layers adds up to reasonable levels.

Shortly before this book was finished, the Livermore team working on unclassified X-ray lasers showed that such reflective cavity mirrors could raise output power from an X-ray laser plasma. However, amplification was limited by the short duration of the population inversion in the plasma. The plasma stayed excited for about 500 picoseconds, but it took the X rays 275 picoseconds to travel the 8-centimeter distance between the mirrors, so once the signal was generated it could pass only once through the cavity while the population inversion could amplify it.

Mirror reflectivities remain limited. The 30-layer rear cavity mirror in the Livermore experiment reflected only 20% of the incident X rays. The high absorption that accompanies that low reflectivity could make it hard for such mirrors to withstand high power levels for more than one or two reflections.

Smaller X-Ray Lasers

X-ray lasers will never become practical for most applications if they require pump power from a nuclear explosion or a giant laser. A few researchers are looking at ways to "scale" X-ray lasers down to smaller sizes. They hope that moving to slightly longer wavelengths will relax pump power demands enough to allow use of modest-power pump lasers. Peter Hagelstein of the Massachusetts Institute of Technology envisions a laboratory-scale X-ray laser system that would cost about $250,000.

X-Ray Laser Applications

What could one do with an X-ray laser—besides using it as a science-fictional weapon? Coherent, collimated beams of X rays could be invaluable in studying the structure of matter and even living organisms. X-ray holograms would permit three-dimensional structural studies of living cells—perhaps even of images recorded while the cells were alive. (The X-ray laser pulse would kill them, however.) Coherent X-ray beams also might be useful for other applications including processing of semiconductors. Efforts have begun only recently to explore these exciting possibilities.

OTHER NOVEL LASER CONCEPTS

A few other laser ideas also do not fit into our standard laser categories. We'll mention them briefly here, to show some other ideas that exist in the research laboratory.

Gamma-Ray Lasers

Gamma rays have even shorter wavelength than X rays. However, they, too, usually are considered to come from a distinct source. Physicists usually define gamma rays as photons produced by transitions between energy levels of the atomic nucleus. These transitions are more energetic than X-ray transitions, so gamma rays have shorter wavelengths.

The shorter wavelengths might make gamma-ray lasers seem even harder to make than X-ray lasers. However, gamma-ray transitions have one big potential advantage—some upper levels have quite long lifetimes, as metastable nuclei. This avoids the need for tremendous pump energies to produce a population inversion as in an X-ray laser.

The story isn't that simple, of course, and some truly formidable obstacles remain. Nonetheless, gamma-ray lasers are at least an interesting theoretical possibility.

Nuclear-Pumped Lasers

The bomb-driven X-ray laser we described earlier is the most dramatic example of a nuclear-powered laser. There also has been a longer-running and much more public effort to tap the energy of more controlled nuclear reactions.

The essential idea is to excite the laser medium with energy from a nuclear chain reaction. Neutrons or fission fragments produced by a pulsed fission reactor could transfer some energy to a gas laser medium, producing a population inversion and laser action. Two 1974 experiments proved that the concept was feasible. However, power levels have never reached the levels that developers hoped for. Research is continuing, with one goal being the use of such reactor-driven lasers in space.

What Have We Learned?

- The lasers most broadly tunable in wavelength are optically pumped organic dyes in liquid solutions. Their major use is to generate tunable-wavelength visible light for scientific research.

- Dye lasers are versatile but complex.

- The broad gain bandwidth of dye lasers lets them generate exceptionally short pulses.

- Dye laser cavities are designed to allow wavelength tuning. They include a prism, diffraction grating, or other wavelength-selective element.

- Dyes have gain bandwidths of 10 to 70 nm, but even without wavelength selection, natural line-narrowing will limit linewidth to a few nanometers.

- A ring cavity limits dye lasers to narrower bandwidth than other cavities.

- Wavelengths emitted by dyes depend on their chemical composition, the pump wavelength, and the duration of pumping, as well as on cavity tuning.

- Dyes decompose gradually during use. Most must be dissolved in organic solvents.

- Dye lasers can be pumped by linear or coaxial flashlamps, pulsed excimer, neodymium, copper-vapor, or nitrogen lasers, or argon or krypton ion lasers.

- Continuous-wave dye lasers can emit light with extremely narrow bandwidth.

- Dye-laser harmonics can be tuned in wavelength by tuning the laser's output.

- Free-electron lasers extract energy from a beam of electrons passing through a wiggler magnet.

- Free-electron lasers can be tuned across a broad range of wavelengths by changing the period of the wiggler field or the electron energy.

- The wiggler magnet field must vary sinusoidally, but it need not come from permanent magnets. It can come from an electromagnetic wave.

- Tapering the wiggler field can increase free-electron laser efficiency.

- Free-electron lasers draw on well-established technology for particle accelerators.

- Free-electron lasers operate in two regimes. At short wavelengths, interactions are between individual particles; at long wavelengths the interactions are collective.

- Different cavities are needed for high-gain Raman and low-gain Compton free-electron lasers.

- The high efficiency of free-electron lasers may allow very high power levels.

- All the potential of free-electron lasers has yet to be realized. Important obstacles remain to producing high powers at short wavelengths.

- X rays are produced by inner-shell electronic transitions. They are ionizing radiation that can strip electrons from atoms or molecules.

- It takes extremely high peak powers to produce an X-ray population inversion.

- Nuclear explosions can provide the energy needed to power X-ray lasers. This highly classified approach is a "Star Wars" program.

- Short, extremely high-power pulses from visible lasers can drive X-ray lasers.

- Reflectivity is much lower at X-ray wavelengths than in the visible. The best mirrors reflect only about 20% of incident X rays.

- Some researchers are working on laboratory-sized X-ray lasers.

- Gamma-ray lasers would operate on transitions of atomic nuclei.

- Energy from controlled nuclear fission can power a laser.

WHAT'S NEXT?

In the next two chapters we will explore the applications of lasers. Chapter 10 covers the uses of low-power lasers. Chapter 11 covers uses of high-power lasers.

9

Quiz for Chapter 9

1. What is the shortest pulse that can be generated from a dye laser with bandwidth of 10 gigahertz?
   a. 44.1 nanoseconds
   b. 4.41 nanoseconds
   c. 441 picoseconds
   d. 44.1 picoseconds
   e. 4.41 picoseconds

2. How does a diffraction grating select a particular wavelength to oscillate in a laser cavity?
   a. It diffracts only one wavelength back in the right direction to oscillate with the other cavity mirror
   b. It increases losses at other wavelengths by diffracting them out of the laser cavity
   c. It reflects only one wavelength and absorbs the rest
   d. A and B
   e. All the above

3. What property gives organic dye lasers their wavelength tunability?
   a. The many vibrational sublevels of electronic transitions
   b. The use of organic liquids as solvents
   c. The high gain of the optically pumped dyes
   d. Photodissociation of the dyes under intense pump light
   e. The liquid nature of the laser medium

4. Which of the following pump sources is used in the lowest-cost dye lasers?
   a. Excimer lasers
   b. Nitrogen lasers
   c. Continuous-wave argon lasers
   d. Third harmonic of Nd-YAG
   e. Electric discharge

5. An excimer-pumped dye laser generates pulses 10 nanoseconds long. What is the minimum bandwidth the pulses can have?
   a. 4.41 gigahertz
   b. 441 megahertz
   c. 44.1 megahertz
   d. 0.4 nm
   e. None of the above

6. What is the wavelength of a free-electron laser powered by 20-MeV electrons with a wiggler with 5-cm period?
   a. 15.6 micrometers
   b. 56.1 micrometers
   c. 0.156 millimeter
   d. 1.56 cm
   e. 5 cm

7. The first stage of a two-stage free-electron laser is powered by a 5-MeV electron beam and has a wiggler with a 5-cm period. What is its wavelength?
   a. 25.6 micrometers
   b. 56.1 micrometer
   c. 0.196 millimeter
   d. 0.216 millimeter
   e. 2.16 millimeters

8. The second stage of a two-stage free-electron laser uses the electromagnetic field from the first stage in problem 7 as its wiggler magnet field. What is its wavelength if it uses the same electron beam?
   a. 216 micrometers
   b. 15 micrometers
   c. 930 nanometers
   d. 515 nanometers
   e. none of the above

9. What types of transitions produce X rays?
   a. Vibronic transitions
   b. Outer-shell electronic transitions
   c. Nuclear transitions
   d. Inner-shell electronic transitions
   e. Nuclear explosions

10. What is the wavelength of 1000-eV X rays?
    a. 1000 nm
    b. 100 nm
    c. 12.6 nm
    d. 1.26 nm
    e. 0.126 nm

Low-Power Laser Applications

ABOUT THIS CHAPTER

Lasers are used for so many applications we can't do them justice in a single chapter. In this chapter, we look at applications that require low levels of laser power, typically well under a watt. Such little power does not have dramatic effect on the objects it strikes, but can make minor changes, such as exposing photographic film. As we will see, there are many different types of low-power laser applications. In some, the laser is little more than a high-performance light bulb, but for others, the special features of laser light—such as coherence or tight beam collimation—are essential.

THE ATTRACTIONS OF LASERS

Theodore Maiman's first laser made headlines in the summer of 1960. As word of the laser spread, it seemed that every scientist and engineer wanted his or her own laser. Many had no clear purpose in mind, but the laser seemed like a neat toy, and curiosity lurks deep inside the souls of scientists and engineers. They built their own lasers, or bought lasers from the handful of little companies that sprang up to make them. Then they set them up and started zapping just about everything that couldn't get away. They informally measured laser power in "gillettes"—how many razor blades a laser pulse could penetrate. It was a fertile time for experiments, new ideas, and accidental discoveries. But like any boom, it busted, and left behind the joke that the laser is "a solution looking for a problem."

Much has changed since the first wave of laser enthusiasm. Today, plenty of jobs demand lasers. In Chapter 11, we will talk about those which use laser power to change something in a major way. In this chapter, we will concentrate on the many uses of low-power lasers.

10

What can low-power lasers do that other light sources can't? For some jobs, they may merely be serving as well-behaved, long-lived light bulbs. However, other jobs may require special properties of laser light, such as the coherence and tight alignment of laser beams. Let's look at the major differences between low-power lasers and other light sources:

1. Lasers produce well-controlled light that can be focused precisely onto a small spot
2. Low-power laser beams, focused tightly in a small spot, can reach a high-power density
3. Laser light travels a well-defined straight line
4. Laser light covers only a narrow wavelength range
5. Lasers generate coherent light
6. Lasers can generate extremely short pulses
7. Diode and helium-neon lasers produce steady powers for tens of thousands of hours or more
8. Diode lasers are very compact
9. Diode lasers can be modulated directly at high speeds by changing drive current

On the other hand, lasers do have some significant limitations that affect how they can be used.

1. Lasers don't emit white light
2. The cheapest lasers are infrared diode lasers
3. The cheapest visible-wavelength lasers emit red light; yellow, green, and blue lasers are much more costly
4. The concentration and collimation of laser light makes it hazardous to look directly into a laser beam

In the remainder of this chapter, we will show how these characteristics combine with other factors to determine how low-power lasers are used. For starters, you might mentally compare a laser to a light bulb. Light bulbs are very good for illuminating rooms, but not for illuminating one tiny spot far from the bulb. On the other hand, a laser is virtually useless for illuminating a room, but it's very good for delivering light to a tiny spot far away.

READING WITH LASERS

Low-power lasers often read printed symbols, such as the striped Universal Product Code (UPC) symbols on food packages, or special type fonts in optical character readers. However, they do not read in the same way that your eye reads. The laser beam illuminates the object being read, but a light detector and an electronic receiver actually decode whatever is being read.

Figure 10-1 is a simplified diagram of the workings of a laser bar-code reader. A beam deflector—here a rotating mirror—scans the laser beam in a pattern across the zone where the bar code will appear. Laser light reflected from that area is focused onto a detector, which generates an electrical signal proportional to the light reaching

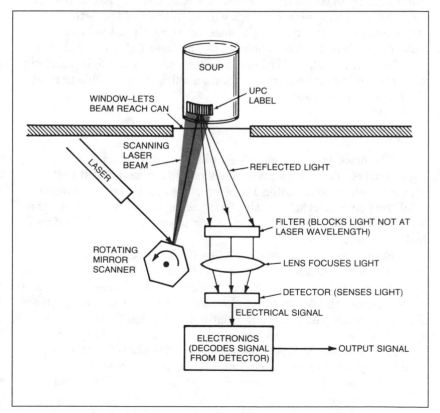

Figure 10-1. A generic laser bar-code reader.

it. The amount of reflected light varies with time as the beam scans across dark and light zones. The detector and its electronic circuits sees those dark and light areas as periods when reflected light is absent or present, respectively, and decodes them as zeros and ones. Thus, the scanner converts a symbol on a piece of paper into an electronic signal.

Our diagram does not show other lighting, but we can't ignore it in the real world. In a supermarket, for example, bright overhead lights shine down onto the checkout counter, illuminating packages at the same time as the laser. However, the laser light is concentrated at one wavelength (632.8 nanometers for a helium-neon gas laser), while room light is distributed throughout the spectrum. You can keep the room light from confusing the detector with a filter that transmits only the laser wavelength, which we put in front of the detector in our diagram. A little room light does get through at the laser wavelength, but it is overwhelmed by the laser light, and does not interfere with reading. Without the filter, most of the light reaching the detector would be from the room, making it impossible to read the bar codes accurately.

Supermarket Scanning

The place you are most likely to see a laser beam is in the supermarket. The laser's presence is not advertised, but it rests under the checkout counter. Its beam comes up through a window to read the bar codes that identify food packages, and you can see the low-power red beam of the helium-neon laser scanning labels if you look carefully.

Supermarket scanners work like the laser scanner in *Figure 10-1*, but there are a few important refinements. The beam-scanning pattern and the bar code are both designed so packages can be moved rapidly across the scanning window without bothering to align them. The scanner can read the code front-to-back or back-to-front; the code itself tells the scanner which is which.

The scanner does not read the price directly from packages supplied by outside companies, such as cans of soup or boxes of cereal. It reads the bar code which identifies a specific product, then looks up the price of that product in a table stored in the store's computer. That price is added to your bill. Many stores print special bar codes for produce and meat that they package. Those bar codes indicate price and weight, which can also be read directly at the checkout.

The symbols used on food packages are the Universal Product Code (UPC), which was designed to be read with a 632.8-nm helium-neon laser. (The symbols also should be readable with the new 670-nm diode lasers.) The wavelength is important because it affects specifications for printing UPC symbols. For the laser to read the bar code correctly, the dark and light stripes must reflect different fractions of light at the helium-neon wavelength, which is all that the detector "sees." It took some food processors a while to learn that it was only the laser wavelength that mattered. For a while, some milk companies printed red-striped bar codes on their white-paper milk cartons. Both the red stripes and the white paper reflected the red laser light strongly, so the symbols were invisible to laser scanners.

Now taken for granted in large food stores, the Universal Product Code has been adapted for many other purposes. Magazines, and even this book, use similar symbols on their covers to speed processing by wholesale distributors. Other retailers use similar codes. The government uses a similar code to identify materials in its inventory. You can buy scanning attachments for personal computers.

Not all scanners contain lasers. Lasers allow scanning from a greater distance, but you also can read codes by running a handheld "wand" containing a red LED over the striped symbol. Such scanners work because they read at the same wavelength as the laser.

Optical Character Readers

Lasers also can be used in optical character readers, which read words printed in special or standard type fonts. In such systems, a tiny laser spot scans across the printed page, and a detector monitors the reflected light. An internal computer uses the pattern of reflected light to build up an internal "image" of the information printed on the page. Then it compares that internal image with printed characters in standard type faces.

Optical character readers are valuable because they let computers read information in human-readable form. Suppose, for example, you wanted to enter information from an old magazine into a printed database. Instead of having to type all the words from the old magazine, you could run the pages (or copies of the pages) through an optical character reader, which would read them for you.

Although optical character readers are far from infallible in interpreting printed words, they are a vast improvement over having to enter every word by hand.

OPTICAL DISKS AND DATA STORAGE

Most of the world's lasers read a different type of information—digitized sound encoded on reflective compact discs. The lasers are tiny inexpensive semiconductor lasers, and they're buried deep inside compact disc players, where you would never find them unless you disassembled the players. Millions of CD players are sold each year, making them impressively successful even by the standards of consumer electronics.

Optical Disk Technology

The basic concept of optical disk technology is shown in *Figure 10-2*. Light from a semiconductor laser is focused onto the surface of a rapidly spinning disk. The surface of the disk is covered with tiny spots, which record bits of data, and have different reflectivity than blank areas on the disk. Laser light reflected from the disk is focused onto a detector, which generates an electrical output. The result is a series of pulses, corresponding to the data recorded on the disk.

Optical storage can cram tremendous amounts of data onto a small surface area. You can focus a diode laser beam to a spot to less than one micrometer across, although data is not recorded quite that close to avoid errors. One optical disk format, called a CD-ROM for *C*ompact *D*isc, *R*ead-*O*nly *M*emory, can store 600 million bytes (4.8 billion bits) of digital data on a 12-centimeter diameter disk—about 85 billion bits per square centimeter. The actual storage density is higher, because not all the disk area is used.

In our example, we only showed the reading of optical data information prerecorded on CDs and CD-ROMs. However, some optical disks have light-sensitive surfaces so you can write as well as read data, by modulating a laser beam pulse which strikes the surface at certain points to store data bits. We'll describe the details below.

Videodisks

Ironically, the thing that originally stimulated interest in optical disks turned out to be a market flop—the home videodisk player. In the 1970s, the Dutch electronics giant N.V. Philips and the American

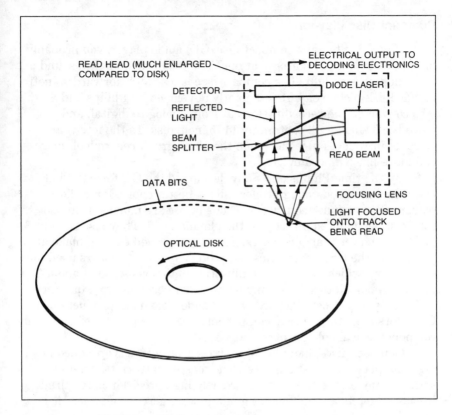

Figure 10-2. Optical disk system.

entertainment firm MCA Inc. both decided to develop home players of prerecorded video programs. They settled on an optical disk approach, with players built around mass-produced helium-neon lasers. Video programs were encoded in analog form on 12-inch (30-centimeter) disks.

Their technology got to market a bit later than planned, but it did work. Unfortunately, videocassette recorders got to the market first, and the public preferred the tape system because it let them record as well as play back programs. Laser videodisk players are still available, but they've never made a big splash on the market. (RCA's laserless videodisk made a bigger market splash, but it proved to be a $500 million fiasco that drowned!) However, videodisk players opened the door for a new generation of optical disk systems.

Compact Disc Players

If you don't own a compact disc (CD) audio player, you probably have at least looked longingly at one. The crisp, clean digital sound is an audiophile's delight. The compact disc is a straightforward spinoff of videodisk technology. It was jointly developed by Philips and Sony, who converted the recording format from analog to digital, and moved to a smaller 12-centimeter (4.75-inch) disk. In the few years since its introduction, it has made the phonograph record look more and more like a dinosaur.

By the time Philips and Sony developed the CD format, diode lasers had matured enough to look like a better choice for audio players than helium-neon lasers. Diode lasers are much smaller, so they can be mounted directly in the playing head. They also require much less power than gas lasers. They have proved easy to make at low cost in the large quantities required. The GaAlAs lasers used for CD players, which emit a few milliwatts continuous wave at about 780 nm, are reported to sell for only a few dollars each in large quantity in Japan, where most CD players are made. The mass production of CD lasers and other optical components has made inexpensive optical components available for other purposes.

Compact discs, like phonograph records, are mass-produced by pressing plastic against a master. The master pattern transfers a spiral of tiny pits to the copy. Extremely high precision is required because the pits are only about a micrometer across, but once that precision is attained, reproduction costs are low. Some market analysts believe it may soon cost only a dollar or two to mass-produce CDs.

CD-ROMs, CD-I, CD-V, and Other Spinoffs

The impressive success of the audio CD encouraged Philips and Sony to modify the format for other applications. The first such modification, and still the most important, is the CD-ROM, or Compact Disc Read Only Memory. Like an audio CD, it uses 12-centimeter disks, but it is digitally encoded with 600 megabytes of computer data instead of music.

The CD-ROM is a very attractive format for publishing large databases and other reference compilations for computer retrieval. One reason is the disk's extremely high capacity; the text from a standard 24-volume encyclopedia fills only about a third of a CD-

ROM. Suitable software allows fast electronic access to information anywhere on the disk. Enter the word "laser," and the electronic encyclopedia will almost immediately display the main entry and cross-references to other entries that mention lasers.

CD-ROM players are similar to those used with conventional CDs, but they don't need the ability to convert the digital data bits into audio format. (However, the need to directly interpret digital data bits makes error correction more important for CD-ROMs than for audio disks, where a single bad bit normally is not detectable.) Likewise, the same technology can produce CD-ROMs as well as CDs, although, again, minimization of errors is more critical for CD-ROMs than for CDs.

The original CD-ROM format is limited to text only. New variants allow some graphics. The CD-I (*Compact Disc-Interactive*) format can handle simple interactive graphics as well as text. The CD-V (*Compact Disc-Video*) format can show short sequences of full-motion video. Other variations in the works promise to make the CD family even more flexible.

Write-Once Optical Storage

You can read data from CDs, CD-ROMs, and other members of the compact disc family, but you can't write onto them. The surfaces of the prerecorded disks are not sensitive to light. If you want to store data on optical disks, you need disks coated with light-sensitive materials that change their reflectivity when exposed to enough light. Then you can store data bits at selected places on the disk by turning the beam off and on at a high enough power level.

The first such materials to be developed are used in the *Write-Once, Read-Many* or "WORM" disks. They let you record data on the surface of an optical disk, but the recording is permanent (like photographic film) and is not erasable like magnetic tape or floppy disks used with computers. Pulses from a diode laser both write and read data on WORM disks. The write pulses have higher power, so they change the disk's surface, but the lower-power read pulses cause no permanent changes. Some computer users have hesitated to accept WORM disks because they are used to working with erasable magnetic media. However, for some purposes, such as backups or archival storage, the inability to erase the WORM disks may make them more attractive.

Like CD-ROMs, WORM disks can store large quantities of data, although in practice the densities usually are not quite as high. Standard recording formats for WORM disks also are quite different from those for CD-ROMs. One quirk in commercializing WORM disks is the choice of a de-facto standard diameter of 13 centimeters, because it is close to the 5 1/4-inch size common for floppy magnetic disks and hard magnetic disks. However, it is the difference in recording formats rather than the 1-cm larger diameter of WORM disks that makes it impractical for a single drive to read both CD-ROM and WORM disks.

Erasable Optical Disks

New types of erasable optical disks can be used like magnetic disks for computer storage—you can write data on them, erase the data, then write new data on the same area. This is possible because they use different materials and different recording techniques than WORM disks.

The actual recording technique is called "magneto-optic" because it relies on a combination of magnetic and optical phenomena. It takes a combination of light from a diode laser and a magnetic field to record a data bit. The disk is coated with a material which can be magnetized if it heated beyond a certain temperature. At lower temperatures it keeps the magnetic field direction it took when heated. A beam from a diode laser heats the coating to the critical temperature, as shown in *Figure 10-3*. Once the coating exceeds that temperature, it becomes magnetized in a direction set by a magnet on the other side of the disk. Heating the material again when it is exposed to a different magnetic field can change the field direction in the coating again, making the data erasable.

Reading is with a lower-power pulse from a semiconductor laser that passes back and forth through the disk coating (which is partly transparent). Magnetization of the disk material affects the polarization in a way that can be detected by looking at the laser pulse reflected from the disk. Thus, although diode lasers write and read the data on the disk, the information actually is stored in magnetic form.

Magneto-optic data storage, like WORM drives and CD-ROMs, allows very high storage density. However, at this writing magneto-optic storage is only starting to reach the market.

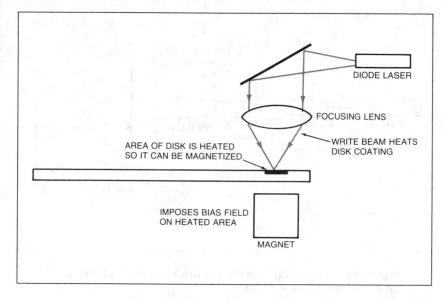

DIODE LASER

FOCUSING LENS

AREA OF DISK IS HEATED
SO IT CAN BE MAGNETIZED

WRITE BEAM HEATS
DISK COATING

IMPOSES BIAS FIELD
ON HEATED AREA

MAGNET

Figure 10-3. Magneto-optic recording relies on heat from a diode laser.

LASER WRITING

Laser beams can write as well as read. The concept is shown schematically in *Figure 10-4*. The laser beam is turned on and off as it scans across a light-sensitive surface. In *Figure 10-4*, when the beam is on it writes a dark spot; when it is off, it does not change the light-sensitive surface. After the laser scans a single line across the surface, the beam or the surface can be moved slightly so the laser can write another row of spots next to the first.

This basic concept appears in many variations. The most common is the laser printer of computer output, now used with many personal computers. Others include laser systems for reproducing printing plates and for setting type.

Laser Computer Printers

Look closely at a laser printer attached to a personal computer, and you'll see a strong resemblance to an office photocopier. There's a good deal of similar technology in the two, and some printers use the

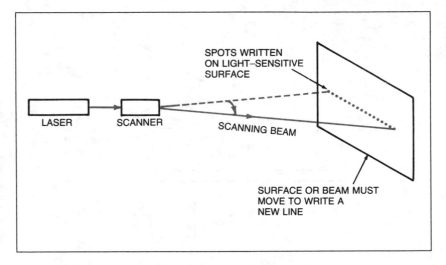

Figure 10-4. Scanning laser beam writes a series of spots on a light-sensitive surface.

same type of rechargeable cartridges used by inexpensive photocopiers.

In an ordinary plain-paper copier, a bright light shines onto the page to be copied, and light reflected from the page is imaged onto a rotating drum coated with light sensitive material. The coating initially carries a static electric charge that is discharged by light reflected from the page. The drum continues rotating after the light strikes it, and the regions that were not illuminated still carry an electric charge so they pick up a dark material called a "toner." The drum then transfers the toner to plain paper, producing a copy of the original page. (The dark toner adheres to the parts of the drum where the dark areas of the original page were imaged.)

A laser printer contains the same drum arrangement, as shown in *Figure 10-5*, but there is no original. Instead, a modulated laser beam scans across the drum's coated surface, writing a pattern of charged and uncharged areas. The rotating drum then picks up the toner, and transfers the image to plain paper as in a conventional copier. Note that the image is made not by copying an original, but by scanning the drum with a laser beam.

With a typical resolution of 300 dots per inch, the output of a laser printer looks almost as good as a typed page. In fact, often the

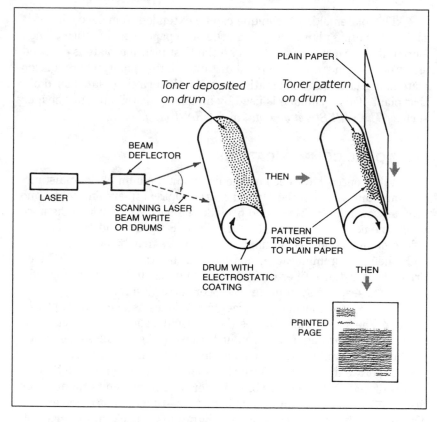

Figure 10-5. Basics of a laser printer.

copy for complete books—including all text and art—are produced using a laser printer.

Laser Platemakers

Another use of laser writing is in making printing plates for newspapers. The newspaper staff enters articles and other material to be printed into a computer, which compiles it into the format needed for publication. Then a high-resolution laser beam scans a light-sensitive printing plate, recording images of the words and pictures to be printed. (Original copies of advertisements typically are inserted by other means.) After the plate is processed, newspaper pages can be printed from it.

This platemaking technique can be extended to make copies at satellite printing plants away from the newspaper's main offices. The electronic information used to write the first printing plate is encoded and stored, then transmitted to a distant printing plant, where a laser platemaker uses that information to make the printing plates used at that plant. This technique is used by nationally distributed newspapers such as *The Wall Street Journal* and *USA Today*.

FIBER-OPTIC COMMUNICATIONS

Fiber optics have become the dominant medium for long-distance telecommunications in the United States. Optical fibers link telephone company switching offices, and provide the backbone for long-distance telephone networks operated by AT&T, GTE Sprint, and MCI Telecommunications. Eventually, optical fibers may bring telephone and other telecommunications services all the way to homes. LEDs can drive short-distance fiber-optic systems, but the light sources for long-distance fiber-optic systems are semiconductor lasers.

Figure 10-6 shows key elements of a long-distance fiber-optic system. The signal that reaches the transmitter generates a current which passes through a semiconductor laser. The signal is a series of digital pulses, and it generates a series of current pulses, which in turn generate light pulses from the laser. The laser emits the light pulses directly into the core of the fiber, which transmits them—over distances up to tens of miles or kilometers—to a distant receiver. There, a light detector converts the light pulses back into a series of electrical pulses.

The semiconductor laser well matches fiber-optic communication requirements. Its small size lets it couple light pulses directly into the tiny cores of optical fibers. It operates on the same voltage and current levels used by conventional semiconductor electronics.

The development of fiber-optic communications has driven much work on semiconductor lasers. Early fiber optics were designed to work at the 800- to 900-nm wavelengths of GaAlAs semiconductor lasers. The discovery that glass optical fibers had lower loss at 1300 and 1550 nm led to development of the InGaAsP lasers which operate at those wavelengths. Output powers of a few milliwatts are adequate for long-distance transmission through optical fibers with typical loss in the 0.5 decibel per kilometer (dB/km) range at 1300 nm and 0.25–0.2 dB/km at 1550 nm. (For more about fiber optics, see the author's book, *Understanding Fiber Optics*, also published by Howard W. Sams & Company.)

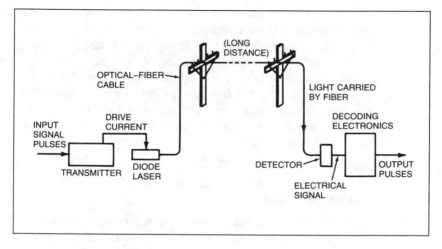

Figure 10-6. Long-distance fiber-optic system.

OPTICAL COMPUTING

Optical devices can manipulate light in ways more complex than merely focusing or bending light rays. They can switch light rays in different directions, block or transmit light, or convert the information carried by a light wave into different forms. This whole range of processes is sometimes called optical computing. Some of them require lasers, some work with laser light or incoherent light, and some only work with incoherent light. Most remain developmental. While we can't go into much detail here, we can list some key concepts and applications:

- *Optical switching*: Like electronic switching, optical switching redirects optical signals, often through optical fibers.

- *Optical interconnection*: Signals can be routed on alternative paths among electronic devices by optical interconnections, which allow routings not possible with electronic conductors. While electronic interconnections typically must be in a circuit plane, optical interconnections can lie outside the circuit plane.

10

- *Analog optical transforms*: The conversion of information presented as a function of time in a time-varying optical signal $f(t)$ into a function of frequency $F(\nu)$ that contains the same information. This operation is called a Fourier transform, and is essential to analyze the frequencies that make up a signal that varies in time. There also are two-dimensional Fourier transforms that convert the distribution of light in the input plane $f(x, y)$ into a distribution of spatial frequencies $F(\nu_x, \nu_y)$, where ν_x, and ν_y are spatial frequencies in the x and y directions, respectively. Spatial frequencies are analogous to time frequencies, but they measure the variation of light in space or position rather than in time.

- *Special analog transforms*: These are used for applications such as conversion of raw data from synthetic-aperture (side-looking) radars into usable images.

- *Vector-vector, matrix-vector, and matrix-matrix operations*: These are difficult to perform with digital electronic computers, but optical systems promise to perform them faster by handling operations on different matrix or vector elements in parallel with arrays of optical elements. Such operations are extremely important in many types of modelling and signal processing.

- *Massively parallel computing*: The inherent ability of optical devices to process many inputs in parallel without interference allows parallel computing for applications other than matrix-vector manipulation. (You can think of this as the ability to pass many rays through a single lens.)

- *"Neural network" models*: The inherently high interconnection possibilities of optics can simulate some functions of the human brain.

The possibilities of optical computing are exciting and nearly endless. Some enthusiasts envision optical computers a thousand times more powerful than today's best supercomputers. To be fair, however, we should warn you that many of these promises have been lurking just out of reach for many years. Nor are optical computers good for simple arithmetic. As one veteran researcher quipped, "optical computers can do things you would never dream of, but they can't balance your checkbook."

LINEAR MEASUREMENTS

Laser beams have proved invaluable for many types of measurement. The two most important involve the use of laser beams as straight lines, and of laser wavelengths as measurement units.

Laser Beams for Alignment

The use of a laser beam to draw a straight line may sound trivial or even mundane, but it is important in the construction industry, surveying, and agriculture. Laser instruments for this purpose use the visible red beam from a low-power helium-neon laser. They have simplified some surveying tasks, and also aid in construction and agriculture.

One of the more important construction applications is in lining up the mounts for suspended ceilings or partitions. A laser is mounted in a tripod, with the beam directed up into a prism that bends the beam so it emerges out the side. The prism rotates in a full circle, sweeping the beam around the walls. Adjust this laser plane generator properly, and it sweeps the laser spot around the walls at the same height. Construction workers mount the hangers for suspended ceilings at this level, so they are all even around a room. Turn the laser plane generator 90°, and it can mark the places for partitions.

In agriculture, laser beams define the gradients of irrigated fields. The slope should be large enough that the water doesn't form puddles, but slow enough that the water doesn't run off too fast. A tripod-mounted laser can draw a straight line at the desired angle (for example, at 1° from the horizontal). Then the farmer can mount a sensor on his grading equipment to automatically keep the blade at the right height as it moves around the field.

Measuring Distance by Counting Waves

The wavelength of light is a very convenient unit for measuring small distances. The trick is to use the phenomenon of the interference of light waves that we described much earlier. Interferometric techniques can make very exact measurements of the distance between two points.

To understand the basic idea, we'll look at a simple example of how interference can let us measure a change in distance. Suppose we start with the arrangement shown in *Figure 10-7A*. Light from a laser is split into two beams with a beamsplitter, then directed at a pair of mirrors. The mirrors reflect the laser light back to interfere with each other at a detector just in front of the laser. Initially, the reflected waves are in phase when they meet at the detector, so the detector sees a bright spot.

Figure 10-7B shows what happens if you move one mirror just one-quarter wavelength further from the laser. This makes that reflected wave's path one-half wave longer than it was, so it is out of phase with the other reflected wave. The two wave amplitudes add to zero, producing a dark spot at the detector. Keep moving the mirror, and the detector will see another light spot, then a dark spot, and so on. If you hook the detector up to electronics that can count the light and dark cycles, it can count the number of quarter-wavelengths the mirror moves.

Because the wavelength of light is small, the measurements are quite precise. One-quarter of the 632.8-nm wavelength of the helium-neon laser is 158.2 nm, so if the mirror was moved one micrometer—a thousandth of a millimeter—it would go through just over three full light-dark-light cycles, each corresponding to a half wavelength.

In *Figure 10-7*, we show the measurement of distance in a straight line. However, interferometry also can be used to measure distance changes across a larger area. In this case, the interference becomes visible as a set of light and dark fringes, tracing paths over the area being studied. Each fringe indicates a change in distance between the two surfaces of one-quarter wavelength.

RANGEFINDING AND LASER RADAR

The same principle used in radar measurements of distance can be used for laser distance measurements. You aim a laser at a target, then fire a short pulse from the laser. Then you measure the time it takes the reflected light to return to your receiver at the laser. The measured time t lets you calculate the distance d from the equation:

$$d = ct/2$$

where,
    c is the speed of light.

You need to divide by two because the light actually travels twice the

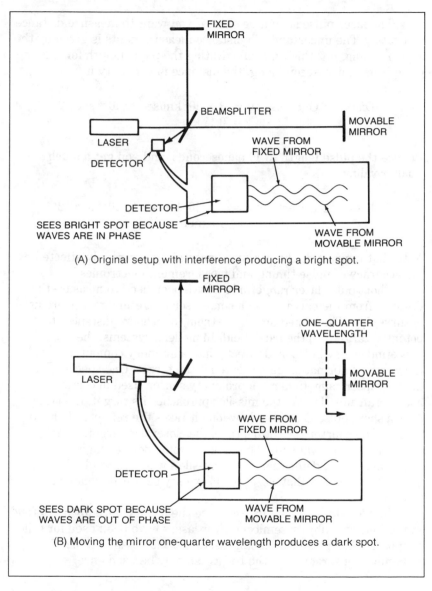

(A) Original setup with interference producing a bright spot.

(B) Moving the mirror one-quarter wavelength produces a dark spot.

Figure 10-7. An interferometer to measure distance.

distance to the target—once on its way there, and the second time on its way back to you.

10

The laser pulse should be short if you want to measure distance accurately. The uncertainty in distance measurements is given by the same formula, but this time substituting the pulse length for t. If the pulse is one microsecond long, the distance uncertainty is

$$\text{Distance Uncertainty} = 3 \times 10^5 \text{ km/sec} \times 10^{-6} \text{ sec} / 2$$
$$= 0.15 \text{ km}$$

Reduce the pulse length to 10 nanoseconds, and you get a much smaller value:

$$\text{Distance Uncertainty} = 3 \times 10^5 \text{ km/sec} \times 10^{-8} \text{ sec} / 2$$
$$= 1.5 \text{ meters}$$

Note that in practice the distance uncertainty is strongly affected by the accuracy of pulse-timing and measurement electronics.

Short-pulse laser rangefinders have been used to measure the distance from the earth to the moon. Laser rangefinders also are used in surveying. The armed forces use them to measure distances to potential targets on the battlefield. In modern systems, the rangefinder may be hooked directly into a gunnery computer, to tell the gunner information he needs to hit the target. One special type of military laser rangefinder is a proximity sensor, used in certain antiaircraft missiles. As the missile approaches its target, the laser fires a short pulse and times how long it takes the reflected light to return. The shorter the return time, the closer the target. The proximity sensor waits for the missile to come near enough to the target plane that detonation of the missile's warhead would destroy the plane. When it gets that close, the proximity sensor detonates the warhead.

Solid-state neodymium lasers are the usual choice for battlefield rangefinders. Ruby or semiconductor lasers have been used for other purposes. Cheap semiconductor lasers are the obvious choice for proximity sensors, where the lasers can only be used once.

OTHER MILITARY TARGETING AIDES

Rangefinders are not the only low-power military lasers on the battlefield. Other types of laser instruments are used to mark targets for precision-guided munitions, or in battle simulation games that were invented years before Laser Tag reached the toy stores.

Laser Target Designators

A laser target designator is an extension on the rangefinder. A designator fires a series of coded pulses at a battlefield target. This series of pulses serves as a unique "mark" that identifies the target to a "smart" bomb or missile that can home in on it.

The smart bomb contains a sensor that looks for the characteristic series of pulses emitted by a target designator. (The coding assures that the bomb isn't misguided by a bright reflection of the sun or some other nontarget source.) The simplest such sensors focus light from the target onto detectors divided into four quadrants. As long as equal amounts of light fall onto each quadrant, the bomb is known to be on course. If one quadrant starts getting more light than the others, the bomb corrects its course to balance the light reaching all the quadrants.

First used in the Vietnam War, laser-guided bombs have proved much more accurate than conventional unguided bombs. However, the use of laser target designators is not without its risks. The soldier holding the designator must keep the target "marked" until the bomb hits it. That means that he must keep the target in sight. Unfortunately, if he can see the target, the target can see him, which can make life rather unpleasant—if not short—if the target contains enemy soldiers who learn they are being marked by a laser designator and can find the soldier with the designator.

Another hazard is eye damage from the laser pulses. Military agencies are not concerned about enemy soldiers they're trying to put out of action, but with the safety of their own soldiers in training exercises. Fortunately, modern soldiers spend more time in training than in real battles.

Laser Battle Simulation

In the 1970s, the Naval Training Equipment Center in Orlando, Florida, worked with military contractors to develop the *M*odular *I*ntegrated *L*aser *E*ngagement *S*ystem, called "Miles." The Miles system equips soldiers with pulsed semiconductor lasers and sensors. The lasers are attached to all kinds of weapons, and each fires a characteristic sequence of pulses. One code indicates the pulses come

10

from a rifle; another a bazooka, and a third denotes heavy artillery. Sensors are strapped on trucks and tanks as well as soldiers.

When the war games start, the soldiers fire laser pulses at each other, and the sensors keep score. A laser-simulated rifle shot can "kill" a soldier. Tanks, however, can only be knocked out by certain types of weapons. (To keep things honest, when the sensors on a tank detect a "kill," they turn off the controls and fire a plume of purple smoke to indicate to everyone on the battlefield that the tank is destroyed.)

Similar laser battle simulation systems are used today by armies around the world. Doubtless it was someone who had seen one of them who invented Laser Tag games, which are based on much the same principle, but don't actually use lasers.

SPECTROSCOPIC MEASUREMENTS

Many of the most sophisticated uses of lasers involve measurement of wavelengths of light absorbed and emitted by various materials, a discipline called "spectroscopy." When we were describing the physical basics underlying lasers, we told how materials absorb or emit light at characteristic wavelengths. Measurements of these wavelengths let us identify and measure the materials that are present.

Fluorescence Spectroscopy

Fluorescence spectroscopy is the measurement of wavelengths at which materials "fluoresce" when they are illuminated by a short-wavelength light source. The emitted wavelengths depend both on the nature of the material and on the wavelength that excited them.

Figure 10-8 shows a simple example of how fluorescence spectroscopy works. Suppose we are looking for samples contaminated with an organic compound that fluoresces at 450 nm when illuminated with 308-nm pulses from a xenon-chloride laser. We can run the samples along a conveyor belt, hitting each one with one or more excimer pulses. We look at the samples through a narrow-band filter that transmits only light at 450 nm. If more than a certain amount of light passes through the filter, we know that the sample is contaminated with the organic compound, and should be rejected. (It might be oil that has gotten into clothing, for example.)

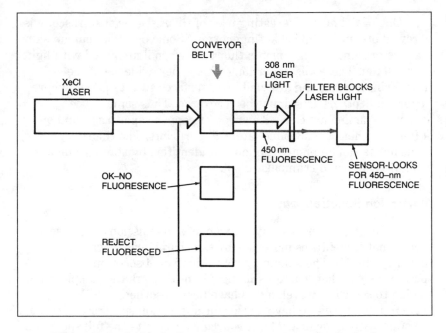

Figure 10-8. A simple fluorescence spectrometer looks for 450-nm emission to detect contamination.

We can do more with fluorescence spectroscopy than our simple example indicates. We can, for example, measure the amount of fluorescence to determine how much material is present. That could let us do a quick chemical test for water pollution.

It's possible to be misled by measurements at only one wavelength because other compounds might fluoresce at the same wavelength. To avoid such problems, we can look simultaneously at several wavelengths. In that case, we could spread out a spectrum across an array of detectors, with each element in the detector array picking up light at a different wavelength. Then we could feed outputs from those detector elements into a computer that would compare the measured spectrum with the spectra of various compounds. For example, fluorescence at 450, 462, 501, and 512 nm might indicate compound A was present, while fluorescence at 450, 480, and 630 nm might indicate compound B was present. Measurement of how much light was emitted at each wavelength would indicate how much of each compound was present.

10

One of the most interesting uses of fluorescence spectroscopy is to reveal otherwise invisible fingerprints. Some oils from human skin that are present in fingerprints fluoresce when illuminated with light near 500 nm. These oils linger long after other evidence of the fingerprint is gone, and are visible on surfaces such as paper, where fingerprints otherwise cannot be recovered. Thus some police agencies examine articles with argon or copper-vapor lasers in hope of finding otherwise-invisible latent fingerprints. The fluorescent fingerprints can be photographed and identified by standard means—and have sent some criminals to jail.

Absorption Spectroscopy

Absorption spectroscopy is analogous to emission spectroscopy except that it identifies materials by the characteristic wavelengths that they absorb. The simplest way to think of absorption spectroscopy is that you are shining a light through the sample, then looking to see what wavelengths have been absorbed.

You don't have to have a laser for absorption spectroscopy. You can start with an ordinary light source which emits the full spectrum of visible light, then spread out the transmitted spectrum to see what wavelengths have been absorbed. However, the laser gives you much finer resolution, so you can identify the precise wavelengths that have been absorbed. With an ordinary light source, you might know only that light was absorbed near 650 nm. A laser could tell you that light was absorbed at two wavelengths in that range, 649.87 and 650.13 nm. Because the laser can concentrate light at a single wavelength, it also can spot absorption lines too weak to see using other techniques.

Dye lasers are the most common types used in absorption spectroscopy. Typically they are "swept" in wavelength, by tuning throughout their emission range. For example, a dye may be tuned from 640 to 670 nm, as shown in *Figure 10-9*. (The laser output could be pulsed or continuous wave, but different measurement instruments would be used for each type of laser.) To measure absorption as a function of wavelength, you would measure the amount of light transmitted throughout the sweep, keeping track of the time. If the entire sweep took 30 seconds, and the tuning rate was uniform in wavelength from the short to the long end of the spectrum, you would know that absorption 10.13 seconds after the start of the sweep was at a wavelength of 650.13 nm.

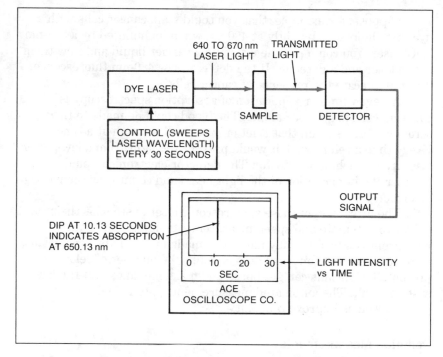

Figure 10-9. Absorption measured by sweeping the dye laser wavelength from 640 to 670 nm every 30 seconds.

Absorption spectroscopy measurements can be made with infrared as well as visible light. The infrared spectrum is often preferred for studies of organic compounds because the absorption bands most useful for identification—arising from molecular vibration and rotation—lie in the infrared. Solid-state color-center lasers and lead-salt semiconductor lasers give a tunable output at those wavelengths, but most infrared absorption spectroscopy still is done with nonlaser sources, because the extremely high resolution possible with lasers is not essential.

Biomedical Diagnostics

One major application of spectroscopic measurements is in biomedical diagnostic instruments. The instruments can use either fluorescence or absorption spectroscopy to look for abnormalities, count cells, or make other types of measurements.

Suppose, for example, that you could stain cancer cells with a dye that fluoresced brightly at 480 nm when illuminated by a 308-nm XeCl laser. You could put the cells into a carrier liquid and flow them past a laser and a detector. If the detector saw 480-nm fluorescence, it would indicate the presence of cancer cells.

An even simpler application of absorption spectroscopy is counting cells in a blood sample. The blood could be made to flow through a tube so thin that it let only one cell through at a time. Every time a cell passed, it would prevent the light from a low-power laser from reaching a detector. The detector electronics would register the interruption of the light beam and count it as the passage of a blood cell.

The attraction of these spectroscopic laser systems is their ability to automate and speed measurements. Many biomedical measurements traditionally have been made manually by technicians, who must look for abnormal cells or count the number of cells present. Technicians can get bored and make mistakes, but the laser system won't. The automated measurement systems help reduce costs as well as improve reliability.

Pollution Measurements

Another important use of spectroscopic laser systems is to detect air or water pollution. For example, you could aim a carbon-dioxide laser at the plume of smoke emitted by a smokestack, and measure the reflected wavelengths. You could tell the quantities of pollutants being emitted by measuring how much light is reflected at certain wavelengths. Thus, environmental agencies can measure concentrations of harmful gases such as sulfur dioxide and nitrogen oxides without having to climb to the top of a smokestack.

The same technology used to detect pollutants near the ground can measure the concentration of ozone (O_3) high in the atmosphere. Ozone is important because it keeps the sun's potentially harmful short-wavelength ultraviolet radiation from reaching the surface of the earth. However, recent measurements have indicated that ozone levels high in the atmosphere are declining, apparently because the ozone is reacting with chlorine released to the atmosphere by the breakdown of chlorinated fluorocarbons. Laser measurements of ozone concentration are helping atmospheric scientists keep track of the problem.

Basic Research

Many uses of spectroscopy remain in the realm of basic research. Laser light, carefully controlled in wavelength, direction and timing, makes an exquisitely sensitive probe of the inner workings of atoms and molecules. Lasers can measure very slight differences between atomic energy levels that are important in evaluating theories of atomic structure. They can measure the very slight differences in energy levels caused by the differing weights of natural and synthetic isotopes. The very slight shifts in wavelength caused by the motion of atoms and molecules themselves can be measured. Lasers have been able to detect the presence of single atoms.

The importance of lasers in exploring the physics of atoms and molecules was recognized by the Nobel Committee in 1981, when it awarded the 1981 Nobel Prize in Physics to Arthur L. Schawlow of Stanford University and Nicolaas Bloembergen of Harvard for their work on laser spectroscopy. (The third person to share in the 1981 Physics Prize, Kai Seigbahn, was cited for research in other types of spectroscopy.)

TIME MEASUREMENTS

The fastest man-made events are pulses of light. The current speed record is six femtoseconds—6×10^{-15} second. That ultrashort pulse of visible light, generated at AT&T Bell Laboratories, was only about three wavelengths long. While the shortest pulses remain hard to generate, pulses lasting about 100 femtoseconds now are comparatively routine, and can be used in many types of research.

What can you do with such ultrashort pulses? They can help measure how fast chemical and physical processes occur. By combining time-resolved measurements and absorption spectroscopy, you can follow progress of a chemical reaction by taking "snapshots" of the composition every 100 femtoseconds. That technique can give valuable information on the speed of chemical reactions, and can identify intermediate compounds that are formed during the reaction.

Fast laser pulses can also measure electrical processes, such as changes in conduction in semiconductor devices. One recent development is an ultrafast laser technique that measures the electric field on the surface of integrated circuits with speed a thousand times faster than the best sampling oscilloscopes. This lets scientists time events in electronic circuits that are too fast to measure with conventional techniques.

HOLOGRAPHY

Virtually all holograms are made with lasers, but few people are aware that holography was invented 40 years ago, more than a dozen years before the laser. Hungarian-born engineer Dennis Gabor, working in Britain after World War II, originally conceived of holography as a way to improve resolution of the electron microscope. His goal was to record and reconstruct not just the amplitude of a wavefront (as in a photograph), but also its phase—information lost in normal photography. That information should allow reconstruction of the entire light wave.

Gabor wasn't thinking of recording three-dimensional images. His original holograms were of flat two-dimensional transparencies. It wasn't until 15 years later, when Emmett Leith and Juris Upatienks started making laser holograms, that anyone realized the possibility for three-dimensional holograms.

Coherence is the property of laser light that is important for making holograms. If you're going to make a hologram of a three-dimensional object, the light you use must have a coherence length at least equal to the dimensions of the object. Other light isn't coherent over such distances, but laser light is. This lets you record both the amplitude and phase of the light reflected from the entire object. Being able to record the wavefront, in turn, lets you reconstruct a three-dimensional image of the original object.

The process of recording a hologram is shown in *Figure 10-10A*. Light from a laser (usually a low-power helium-neon laser) is split into two beams by a beamsplitter. One beam forms the object beam, which is reflected from the object. The other forms the reference beam, which follows a separate path, and is combined with the reflected object beam to record the hologram.

In *Figure 10-10A*, we show the hologram being recorded on a photographic plate. The plate actually records an intensity pattern—the interference pattern generated by combining the object beam and the reference beam. This interference pattern contains information on the phase of the object beam. All you have to do is create another reference beam (possible because you know its characteristics), and shine it at the hologram. Scattering of the reference beam from the

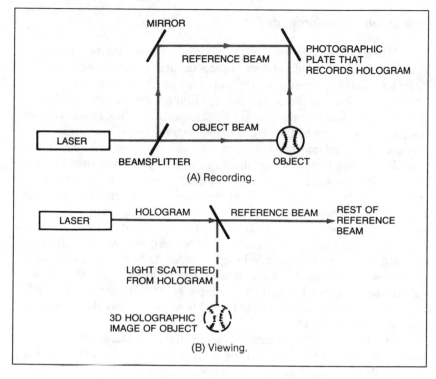

Figure 10-10. Recording and viewing a hologram.

interference pattern recorded on the hologram generates a light wave with the same phase and intensity distribution as the original object beam. The result is a three-dimensional image floating in space, roughly the same size as the original object, as we show in *Figure 10-10B*.

The first holograms had to be viewed as well as created by laser light, as we show in *Figure 10-10*. However, later advances made it possible to view laser-created holograms in white light. Careful choice of the way the hologram is recorded and viewed lets white light serve as the reference beam. You can view so-called "rainbow" holograms in any color of the spectrum; the appearance depends on the viewing angle. These rainbow holograms can be mass-produced by stamping in metallized plastic, so they can be made cheap enough to use on the covers of special issues of magazines including *National Geographic*, *Lasers & Applications*, and *New Scientist*.

10

Holographic Interferometry

Holography can be combined with interferometry to make sensitive measurements of how objects deform under stress. The basic idea is fairly simple, and requires a pulsed laser to take holograms. (Ruby is the usual choice, although frequency-doubled Nd-YAG lasers also can be used.) First, you take a hologram without applying any stress to the object. Then you apply an increment of stress and record another hologram on the same photographic plate.

When you view the combined hologram, you see a single image of the object crossed with bright and dark lines. These lines are the interference fringes that we described earlier. They measure how far the object's surface moved between exposures. The amount and uniformity of change is important. Suppose you tested an aircraft tire before and after applying pressure. If the changes are uniform, they indicate that the tire expanded slightly over its entire surface when air pressure increased. That is what should happen if the tire is sound. On the other hand, if the changes seem concentrated in a small area, they usually betray a flaw less able to withstand strong force than the rest of the tire.

Holographic interferometry has become the standard way to assess the quality of aircraft tires, and it also is used to assess the quality of many other high-performance components.

Holographic Optical Elements

We have seen how holograms can reconstruct three-dimensional images. However, there's another way to view holograms—as optical elements. From that viewpoint, the diffraction pattern recorded in the hologram scatters or bends light in a controllable way. By using this approach, you can design a hologram that functions as a sophisticated lens, redirecting light by scattering it. Typically, holographic optical elements are designed by computer programs.

Holographic optical elements are not as efficient in redirecting light as lenses or mirrors, but they can do things that are difficult or impossible with other types of optics. One example is a large holographic lens used in military aircraft displays. The lens projects an image of the control panel in the air in front of the pilot, forming what is called a "head up" display. This lets the pilot avoid looking down at his dials and controls in the midst of a battle. The large area and short focal length make it impractical to design the lens as anything but a holographic optical element.

ART & ENTERTAINMENT

We have yet to cover the most striking uses of low-power visible lasers—in displays and light shows. Millions of people around the world have seen brightly colored laser beams light up the night sky, a theater, or a planetarium.

The concept is fairly simple. Just bounce beams from visible lasers off moving mirrors. The mirrors scan the laser beams in patterns in the air or on the screen. The scanning patterns can be controlled by music, by a computer program, or by someone operating the lasers (who becomes the optical analog of a musician). It is fascinating to watch the patterns unfold, and the display is safe as long as the beam is kept out of people's eyes. In practice, the key to a good laser display is imaginative control of the scanning system.

Many types of visible lasers have been used in displays. A few light shows have used dye lasers, but typically argon- and krypton-ion lasers, which generate powers of a watt or more at several visible wavelengths, are used. The argon-ion lines generate green and blue lines. Krypton has a strong red line, and also can produce yellow—which is a hard color to find in laser light. The different wavelengths can come from separate lasers, or in some cases from a single "mixed gas" laser which contains both argon and krypton and emits on lines of both.

10

What Have We Learned?

- Laser light illuminates words and codes so they can be read by special detectors and decoding electronics.

- UPC symbols are designed to be read at the 632.8-nm helium-neon laser wavelength.

- Laser-based optical character readers can decode printed words.

- Most lasers in the world are used in compact disc audio players, which play digital audio from 12-cm disks using diode lasers.

- Optical disk technology was initially developed for home video playback. Videodisks flopped, but the technology survived.

- The mass production of CD lasers and other optical components has made inexpensive optical components available for other purposes.

- Other formats have been adapted from CDs. The CD-ROM format can store 600 megabytes of digital data for computer use.

- Lasers can store data by writing on light-sensitive disks. The first such disks developed allow writing once, but not erasing.

- Erasable optical disks are based on magneto-optic materials. Data is read and written optically, but stored magnetically.

- A scanning laser beam can write text or images on a light-sensitive surface. Laser computer printers write with a diode laser on a photocopier drum.

- Laser platemakers reproduce copies of newspaper printing plates at remote locations.

- Diode lasers are the light sources used in fiber-optic systems for long-distance telephone communications.

- Optics can switch and manipulate light to process signals and perform computing functions.

- Lasers can draw straight lines and planes to align machinery and construction equipment.

- Interferometry uses interference effects for precise measurements of short distances in units of the wavelength of light.

- Laser radars measure distance by timing the round trip of a reflected pulse.

- A major use of laser rangefinders is pinpointing targets on the battlefield.

- Laser target designators fire coded pulses to identify a target to sensors on homing bomb or missile.

- Laser pulses can be used to simulate weapons fire in war games.

- Spectroscopy is the study of the wavelengths absorbed and emitted by materials.

- Some materials can be identified by the wavelengths at which they fluoresce when illuminated by a shorter wavelength.

- Laser-induced fluorescence can reveal otherwise invisible fingerprints.

- Biomedical diagnostics are a major use of spectroscopic measurements.

- Spectroscopic laser systems can detect air or water pollution.

- Laser spectroscopy is particularly important for basic research.

- 2-D holograms were made before lasers were available. The long coherence length of lasers made 3-D holograms possible.

- Holographic interferometry can measure deformation of objects under stress.

- Holograms can serve as optical elements.

- Bright scanning laser beams make striking displays and works of art.

WHAT'S NEXT?

In our final chapter, we will study the uses of high-power laser beams.

10

Quiz for Chapter 10

1. Which of the following are differences between laser light and light from other sources?
   a. Laser light has much longer coherence length
   b. Laser light can be focused onto a small spot
   c. Laser light contains only a narrow range of wavelengths
   d. Lasers can generate extremely short pulses
   e. All of the above

2. Which of the following is an important limitation of lasers?
   a. Laser light is invisible
   b. Light rays from a laser are parallel
   c. Inexpensive lasers do not have wavelengths shorter than the red end of the visible spectrum
   d. Lasers do not have long lifetimes
   e. Laser light interferes with itself

3. What type of light sources must be used to read the Universal Product Code on food packages?
   a. Red light sources
   b. Helium-neon lasers
   c. Fluorescent lights
   d. Any laser
   e. Diode lasers

4. What was the first type of optical disk developed?
   a. CD-ROM
   b. Videodisk
   c. Compact disc
   d. Write-Once, Read-Many (WORM)
   e. Erasable

5. What are the differences between WORM and CD-ROM disks?
   a. Data can be written once on WORM disks but not at all on CD-ROMs
   b. WORM disks are 13 cm in diameter, CD-ROMs are 12 cm
   c. Standard recording formats are different
   d. A and B
   e. A, B, and C

6. Which of the following is not a major application for semiconductor diode lasers?
   a. Fiber-optic communication systems
   b. Optical disk data storage
   c. Compact disc players
   d. Supermarket scanners
   e. Laser printers

7. Which of the following is not a promising application for optical computing and processing?
   a. Interconnection of chips
   b. Balancing your checkbook
   c. Matrix-vector operations
   d. Fourier transforms
   e. Processing synthetic-aperture radar signals

8. An interferometer that uses the 632.8-nm light from a helium-neon laser has measured motion of a mirror in an arrangement like *Figure 10-7* as 15 pairs of dark-light fringes. What distance does this equal?
   a. 632.8 nm
   b. 2.37 micrometers
   c. 4.75 micrometers
   d. 7.12 micrometers
   e. 9.49 micrometers

9. A pulse of light returns to a laser rangefinder 100 microseconds after it is emitted. How far away is the object being measured?
   a. 30 meters
   b. 1 km
   c. 15 km
   d. 30 km
   e. None of the above

10. How is laser spectroscopy used?
    a. In basic research
    b. To measure concentrations of pollutants
    c. To count blood cells
    d. All the above
    e. None of the above

High-Power Laser Applications

ABOUT THIS CHAPTER

In Chapter 10 we described the major uses of low-power lasers, loosely defined as those which don't make major changes to objects. In this chapter, we will talk about major applications of high-power lasers. The major applications practical today include cutting, welding, drilling, and marking in industry, some specialized tasks in the manufacture of electronic components, and some medical treatments. Several other applications for high-power lasers remain in research and development, including producing chemical reactions, purification of isotopes, laser weapons, and thermonuclear fusion.

HIGH- VS LOW-POWER LASER APPLICATIONS

The line we have drawn between high- and low-power laser applications can be a thin one. What we call high-power applications are those which cause major changes to the object the laser beam strikes. What do we mean by "major"? Cutting and welding clearly are major changes, whether the cutting is of sheet steel or surgical removal of tissue. The marking of serial numbers and product codes on components is closer to the line, but we tend to consider that a subcategory of industrial materials working.

On the other hand, we put applications that cause changes in light-sensitive materials on the other side of the line. These include things like laser printers and optical data storage. Part of our rationale is that some of the same changes can be made by ordinary light—like the exposure of photographic film. Part is that some of these applications are closely related to other low-power applications—for example, optical data storage by writing on disks is one of a family of optical disk applications that logically belong together. A final part of the rationale is the need to make some separation lest these final two chapters merge into a single monstrous and unmanageable whole.

In reading the last chapter, you may have noted the diversity of low-power laser applications. There is less evident diversity in the use of high-power lasers. Many are variations on the basic theme of laser materials working—heating an object to temperatures sufficient to vaporize its surface. From this standpoint, many medical applications are merely specialized materials working (a technically accurate view, if a bit callous to the patient), and laser weapons are merely intended to perform materials working on unfriendly objects.

You will also note that some of the more interesting high-power laser applications remain in development, even after many years of exploration. Laser weapons and fusion are prime examples. Despite the expenditure of billions of dollars on both technologies, the prospects for the practical use of either remain uncertain. In large part, this represents the massive scales of the undertakings, both in time and magnitude of effort.

Finally, all uses of high-power lasers have benefited in some way from military support. Military basic-research programs helped develop lasers of the power levels needed for materials working and medicine. As we will see later, other programs have been more or less directly supported by military programs. Most of that military support has been stimulated by interest in lasers as weapons.

ATTRACTIONS OF HIGH-POWER LASERS

In Chapter 10, we listed the major attractions of low-power lasers. Let's take another look at the attractions of lasers, this time concentrating on those important for high-power applications:

1. Lasers produce well-controlled light that can be focused precisely onto a small spot.
2. High-power laser beams can produce very high power densities when focused onto small areas.
3. Lasers do not apply physical force to the object the beam strikes, and thus do not move it.
4. Lasers can generate light in short pulses, thus allowing very high power levels to be reached for short intervals.
5. Lasers can be controlled directly by robotic systems.

6. Different laser wavelengths are absorbed by different materials.

7. Laser light travels a well-defined straight line

This list is not identical with the one in Chapter 10, but some points are quite similar. On the other hand, the list of disadvantages of high-power lasers is rather different:

1. Lasers have high capital cost.

2. Many materials strongly reflect the wavelengths emitted by the most powerful lasers.

3. Laser light cannot cut deeply into materials.

4. In some cases, laser light can be hard to control completely.

5. Extremely powerful beams are hard to deliver to targets.

In the remainder of this chapter, we will look systematically at the major uses of high-power lasers and their limitations.

MATERIALS WORKING

Materials working is the broad field of cutting, welding, drilling, and otherwise modifying industrial materials, including both metals and nonmetals. The 1960s engineers who played at measuring laser power in gillettes were—at least in some primitive sense—exploring one aspect of materials working (drilling holes in razor blades).

It's instructive to look at some of the earliest successes of lasers in materials working because they highlight how to take advantage of the properties of laser beams. We'll look at two in particular before moving on to look at materials working in general—drilling holes in diamond dies for drawing wire, and drilling holes in baby-bottle nipples.

Drilling Diamond Dies

An important way of manufacturing thin metal wires is by pulling the metal through tiny holes in diamond "dies." Diamond is an excellent die material because it is the hardest substance known. It also conducts heat well, and thus can dissipate the heat generated by friction when pulling the metal through the hole.

The problem is that the hardness of diamond also makes it an extremely difficult material to drill. Diamond bits get dull when trying to drill holes in diamond, and other materials do even worse. Laser light, however, does not get dull. If a laser beam is focused onto a tiny spot in the diamond, it can remove the carbon atoms—heating them so they combine with oxygen in air or merely evaporate. Deposit enough laser energy, and you can make a hole through the diamond.

Engineers discovered that lasers could drill holes in diamond dies during the 1960s, and ever since this has been a standard example of laser applications. It is not a high-volume application, but it is one where there are no attractive alternatives.

Drilling Baby-Bottle Nipples

The problems in drilling holes in baby-bottle nipples are quite different than those in drilling diamond. The rubber used for nipples is very soft and flexible, which is what makes it hard to drill holes in the material. For milk to flow properly, the holes in the nipples must be small. The best way to make such holes mechanically is with tiny wire pins, but the rubber bends, and the fine pins can be caught and broken by the flexible nipple material. The task is something like trying to punch holes in Jell-O with a thin soda straw.

A laser beam is an excellent alternative because it doesn't contact the nipple and thus doesn't apply any force to it. The laser pulse simply burns a tiny hole through the nipple, without any chance for anything to get stuck. Like drilling holes in diamond dies, this laser application has been around for many years and doesn't require many lasers—but it is nonetheless the best solution to the problem.

Noncontact Processing

As we have indicated, one important factor in drilling both diamond dies and baby-bottle nipples is that laser drilling is a noncontact process. There's no physical laser "bit" that can get dull while drilling diamond. Nor is there any laser probe that applies force that might distort the baby-bottle nipple or be caught while drilling the hole. Avoiding the need to bring a drill bit against the object also speeds processing.

The noncontact nature of laser processing is important for many types of laser drilling, cutting, and welding. Interestingly, many of the materials most often machined by lasers tend to fall into categories similar to diamond and rubber—either very soft or very hard. Titanium, a light metal used in military aircraft, is too hard to cut easily with conventional saws, but can be cut readily with a carbon-dioxide laser. Other advanced alloys and hard materials also are often cut with lasers. On the other extreme are some rubbers, plastics and other soft materials that usually are cut with lasers.

Wavelength Effects and Energy Deposition

Our examples haven't said much about the fundamental interaction that drives laser materials working—the transfer of energy from the laser beam to the object. The actual process is complex, but you should have at least a general understanding of what happens to appreciate the nature of laser materials working.

We saw earlier that all surfaces reflect, transmit, and absorb some energy. If we want laser energy to do something to an object, it is energy absorption that is most important. Reflected energy doesn't do any good, because it is not transferred to the object. As we indicated earlier, the amount of energy reflected or absorbed depends on wavelength. Differences in absorption at different wavelengths can be quite large, and can make certain laser applications impractical.

Table 11-1 lists reflectivity of some common metals and a few other materials at various laser wavelengths. Note that metals become increasingly reflective at longer wavelengths, while skin and paint absorb more light at longer wavelengths. The differences in absorption at the 1.06-micrometer neodymium wavelength and the 10.6-micrometer carbon-dioxide wavelength are dramatic for some metals. Other metals have discouragingly high reflectivity at both wavelengths. Aluminum is strongly reflective at both wavelengths. Worse yet, it's soft and easy to cut with conventional saws and blades. Titanium, on the other hand, absorbs more strongly at 10.6 micrometers than other materials, and is so hard it is almost impossible to cut with a conventional saw.

We should note that low absorption at 10.6 μm does not make it impossible to cut metals with carbon-dioxide lasers. Although only a small fraction of the laser light is absorbed, the carbon-dioxide laser can deliver much more power than Nd-YAG lasers, offsetting the higher absorption at the shorter wavelength.

Surface absorption does not tell the whole story, however. Once the surface absorbs some energy, the material starts to melt, then vaporize. For most materials, the absorption increases with

Table 11-1. Percent Absorption of Materials at Important Laser Wavelengths

Material	Argon Laser 500 nm	Ruby Laser 694 nm	ND-YAG 1064 nm	CO$_2$ 10.6 μm
Aluminum	9	11	8	1.9
Copper	56	17	10	1.5
Human skin (dark)	88	65	60	95
Human skin (light)	57	35	50	95
Iron	68	64	–	3.5
Nickel	40	32	26	3
Sea water	*	*	*	90
Titanium	48	45	42	8
White paint	30	20	10	90

*Sea water transmits most light at these wavelengths that is not reflected at the surface

temperature, so the first bit of heating is the hardest. Moreover, many metals absorb even more light after they have been melted.

However, vaporization can create other problems, by forming a layer of plasma or vapor between the laser and the object, as shown in *Figure 11-1*. This plasma layer can absorb some of the laser energy, blocking it from reaching the object. Sometimes, turbulence produced by heating the object can remove some or all of this plasma. However in other cases the plasma has to be removed to improve energy coupling. The simplest way often is blowing the plasma away.

Another complication comes as the laser beam starts going deep into the material. The deeper the hole, the further the material vaporized by the laser beam must go to get out of the hole. Depending on the depth of the hole and the temperature of the vaporized material, some of it can be deposited in the hole or around it. At best this can reduce the efficiency with which the laser removes material. At worst, it can stop the process altogether.

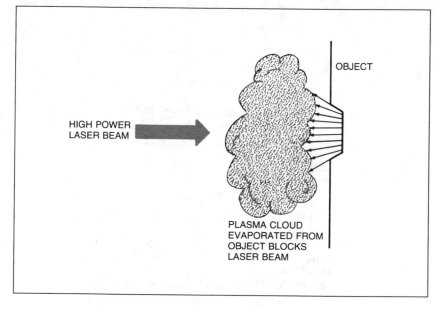

Figure 11-1. Formation of a plasma blocks the laser beam from reaching the object.

The focusing of the laser beam also influences the maximum depth of the laser hole. *Figure 11-2* shows the difference between a lens with short focal length and one with long focal length. The short-focus lens produces a smaller spot on the object, but the beam also spreads out faster beyond the point where the focal spot diameter is smallest. The longer-focus lens does not produce as tight a focal spot, but the beam does not spread out as fast at greater distances, making it possible to drill deeper holes and cut thicker sheets of material.

The Drilling Process

When we talk of drilling, we specifically mean making a single hole through an object. Drilling is done with one or a series of short laser pulses, each of which removes some material, making the hole deeper. The process depends on peak power, and thus works well for short repetitive pulses. (If the pulses are too long, the peak power may not reach high enough levels to vaporize and remove material from the hole.) The number of pulses required depends on factors including wavelength, peak power, the nature of the material, and repetition

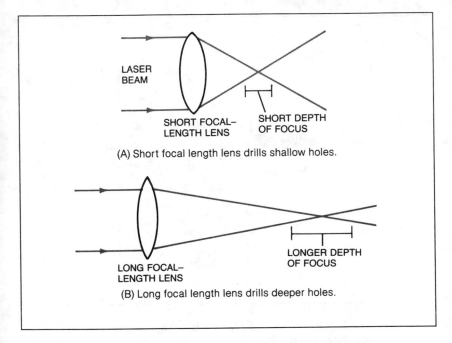

(A) Short focal length lens drills shallow holes.

(B) Long focal length lens drills deeper holes.

Figure 11-2. Effect of focal length on depth of focus. The longer the depth of focus, the deeper the hole that can be drilled.

rate. For most efficient drilling, it may be necessary to change the optical focus to reach into the drilled hole. The laser beam and/or the object are moved to change the spot being drilled. It also is important to make sure that the vaporized material gets out of the hole.

It is simplest to envision drilling one hole at a time with a series of laser pulses. However, in a few applications many holes are drilled at once, sometimes with a single pulse that divided into several lower-power pulses by passing through one or more beamsplitters. One example is the "drilling" of ventilation holes in cigarette papers. A pulse from a carbon-dioxide laser is split into many separate pulses, each of which punches a separate hole in the cigarette paper.

Marking and Scribing

The processes of laser marking and scribing are similar to drilling, but don't require making holes all the way through the material. In marking, the goal is simply to write an indelible message, serial number, or trademark on a component. This can be done by drilling shallow holes without bothering to penetrate deep into the material. The shallower the hole, the faster the job can be done.

Scribing involves drilling somewhat deeper holes, more or less as a perforated line, in a brittle ceramic material. When the ceramic is bent, it breaks along the weakened line scribed by the laser pulses. This technique is most often used in electronics manufacture, a topic described in more detail later.

Cutting Processes

Laser cutting processes resemble those used in drilling with some important exceptions. The motion of the beam and/or the object is continuous, and the beam itself normally is continuous rather than pulsed. You can think of laser cutting as a process something like slicing with a knife, but in this case the knife is a beam of light.

Cutting typically is done with the assistance of a jet of air, oxygen, or dry nitrogen. For nonmetals, the role of the jet is to blow debris away from the cutting zone, and improve the quality of the cut.

With metals, the jet plays a different role. The laser beam heats the metal to a temperature hot enough that it burns as oxygen in the jet passes over it. This process is properly called "laser-assisted cutting," because the oxygen in the jet actually does the cutting.

Welding

Welding may sound similar to cutting, but the two processes differ in fundamental ways. Cutting is the separation of one object into two (or more) pieces. Welding is the joining of two (or more) pieces into a single unit.

Special parameters must be considered in welding, as shown in *Figure 11-3*. In this example, we see two pieces of metal about to be welded together. Note that they must be fitted together precisely, with very little room between them. The laser beam heats the edges of the two plates, causing them to fuse together where they are in contact. If the pieces are not in contact along the entire junction, the weld will be flawed. The laser beam must penetrate the entire depth of the weld to make a proper joint. The heat of welding transforms the metal through the entire depth of the weld, and in a zone extending on both sides of the actual junction. The width of that weld zone depends on the nature of the materials being welded and on the laser energy and scanning speed. The further from the midpoint of the weld, the less the heat has affected the metal.

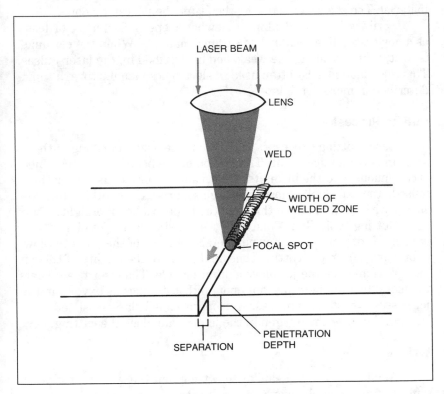

Figure 11-3. Two metal plates being welded with a laser.

The compositions of the pieces being joined together also are very important in welding. The materials do not have to be identical, but metallurgists have learned that some compositions do not bond together well, no matter how thoroughly they are heated, or what technique is used for welding.

Heat Treating

Another materials-working use of high-power lasers is heat treating, a process in which a metal surface is heated to convert the surface layer to another crystalline state. Typically, heat treating converts the surface layer to a crystalline state that is harder and more resistant to wear than the rest of the material. The crystalline form of the metal may be too brittle to use in bulk form, but strong enough if it merely forms a layer on another type of metal crystal.

In laser heat treating, the beam—usually from a carbon-dioxide laser—scans across the surface of the metal. The surface may be covered with a coating to help it absorb the laser energy more efficiently. There are alternative ways to heat-treat metal surfaces, but a laser beam can reach into areas that other processes cannot, such as the insides of engine cylinders.

Materials-Working Power Levels

The laser power levels chosen for particular processes depend upon the type of process, the type of material, its thickness, the laser wavelength, and other factors. Carbon-dioxide laser powers of under a kilowatt generally suffice for most applications involving nonmetals. Multikilowatt carbon-dioxide lasers may be used for cutting or welding thick metal plates (up to an inch or two thick), and for heat treating.

In some applications, you can trade speed with which the beam moves off against laser power. That is, you can use a lower-power laser if you are willing to move the beam more slowly. For some processes, the results of moving a four-kilowatt beam at two meters per second are functionally identical to those of moving a two-kilowatt beam at one meter per second. That is, the same amount of energy is deposited on the area being cut or welded. However, if processing speed is a critical factor, the higher power laser may be preferred.

ELECTRONICS MANUFACTURE

In our discussion of hole drilling, we mentioned how lasers "scribe" ceramic wafers. This is one of several uses of lasers for materials processing in the electronics industry. They differ from the more general materials-processing applications described before mostly in their concentration on precision. Scribing, for example, involves drawing fine lines in brittle ceramic and semiconductor wafers, so they can be broken into precisely sized chips for components.

Wafer scribers and other laser systems built for the electronics industry are customized devices, designed and built for a specific application, not for general-purpose use. They may be built around solid-state neodymium lasers, argon lasers, carbon-dioxide lasers, or in certain cases other types.

Another special-purpose laser system for electronics manufacture is a resistor trimmer. This is a laser built to modify thin-film resistors deposited on hybrid circuits. The resistance of these devices depends on their surface area, which can be modified by removing some surface material with a laser beam. This removal fine-tunes resistance to the desired value.

Other laser systems can repair masks used in making integrated circuits. Suppose, for example, inspection of an IC mask shows a blob of excess material in a critical spot on the circuit. A laser can vaporize the excess material, cleaning up the mask so it can be used to manufacture good electronic components.

Likewise, laser pulses can customize certain types of integrated circuits by making or breaking connections. These special circuits are designed with laser customization in mind, and include extra connections that can be broken or made with a laser beam. If the chip is used for one application, the manufacturer can break one connection; if it is used for another application, the company can break another connection or set of connections. Changing these connections changes the behavior of the integrated circuit. Although many specialized chips have to be altered, the economics of integrated-circuit production favors mass production of many identical chips over small runs of customized chips.

Other potential laser applications in making semiconductor electronics are being studied in the research laboratory. One important example is the deposition of patterned metal films on semiconductors. Recent experiments have shown that lasers can decompose metal-containing gases just above the surfaces of semiconductors or insulators, to produce metal films that conduct electricity well. Researchers are working on extending this technique so they can use lasers to write electrically conductive paths on electronic circuits.

MEDICAL TREATMENT

Physicians weren't far behind engineers when it came to starting experimenting with lasers. The first to use lasers were dermatologists and ophthalmologists (eye specialists), who were used to working with light. However, it was not long before other specialists were looking at potential uses of lasers. Today, important fundamental work is still being done on the interaction between laser light and tissue, but medical laser researchers do have a good understanding of the basic processes involved.

Fundamentals of Laser-Tissue Interactions

We can consider the interaction between laser light and tissue as a problem in light absorption as a function of wavelength. Tissue, like any other material, absorbs light differently at different wavelengths, as you can see in *Table 11-1*. The characteristic wavelengths at which tissue absorbs light depend on its composition, and we can get an idea of how tissue should behave if we look at its major components.

The most important component of tissue is water, which absorbs strongly in much of the infrared region. The absorption is so strong and tissue contains so much water that it isn't a bad approximation to consider tissue as absorbing light like water, especially in the infrared region. At the 10.6-micrometer wavelength of carbon-dioxide lasers, water reflects about 10% of the incident light, and absorbs about 80% in the first 20 micrometer thickness. Absorption is even stronger at 3 and 6 μm, although reflection is somewhat higher at those wavelengths.

The 10.6-μm wavelength has become a favorite for laser surgery because it is readily available and absorbed strongly by water and tissue. Focus a carbon-dioxide laser onto tissue at high enough intensity, and the laser energy will vaporize the cells. The absorption is so strong that only the upper layer of cells will be killed; the lower level of cells will survive with little damage.

The interaction between tissue and carbon-dioxide laser light has another useful aspect. The laser penetrates just deep enough into tissue to seal small blood vessels and stop bleeding. This makes the CO_2 laser an especially valuable tool for surgery in regions rich in blood vessels, such as the female reproductive tract. Water absorption at the 10-micrometer line is not strong enough to let the carbon-dioxide laser cut bone, but surgeons do have other tools they can use. The CO_2 laser's ability to remove thin layers of blood-rich tissues gives it a special slot in the surgeon's collection of tools.

Other lasers have other attractions for medical applications. The

1-micrometer wavelength of the neodymium laser, for example, can penetrate tissue more deeply because it is not strongly absorbed by water. That is not necessarily good, because its greater penetration depth can make it harder for the surgeon to control where the laser beam is going. However, the 1-μm wavelength also can be transmitted by present glass optical fibers, which cannot carry the 10-μm CO_2 wavelength. This lets optical fibers deliver laser light within the body without the need for surgical incisions.

Other visible and near-infrared laser wavelengths can serve other medical purposes. Advances in the state of the art of laser medicine are leading to better criteria for matching lasers to specific treatment needs. If you wanted to target a particular thing in the body—for example, abnormal blood vessels—you would search for wavelengths that the target tissue absorbs much more strongly than surrounding tissue. The idea is to concentrate the laser energy in the tissue you want to remove, while causing minimal side effects to surrounding tissue.

As we will soon see, a growing number of laser medical techniques were developed using this approach. The physician picks the best laser for the job, selecting a desired pulse length as well as wavelength to control energy deposition. The desire to control wavelength has led to development of medical laser systems built around dye lasers, because their wavelengths can be tuned to the best possible value.

Laser Surgery

The surgical removal of tissue, shown conceptually in *Figure 11-4*, is the simplest medical application of lasers to describe. The surgeon scans the beam from a carbon-dioxide laser across the tissue to be removed. Water in the cells absorbs the 10-micrometer beam strongly, vaporizing the top layers of cells, but doing little damage to the underlying tissue. The process also seals off (or "cauterizes") blood vessels smaller than about a millimeter in diameter, effectively stopping bleeding.

Use of carbon-dioxide lasers has become routine for some types of surgery. It is particularly valuable for gynecological surgery, because it can remove abnormal tissue from the surface of blood-rich tissues. For example, lasers can remove abnormal and potentially precancerous cells on the surfaces of the vaginas of women whose

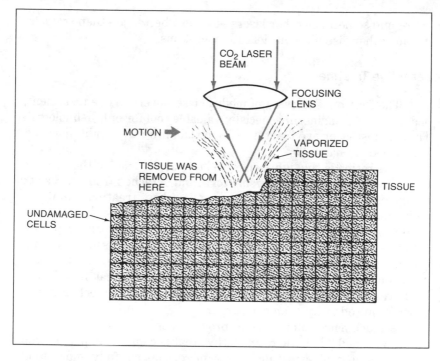

Figure 11-4. A CO₂ laser removes the top layer of cells with minimal damage to those underneath.

mothers took the drug DES while they were pregnant. Lasers can also treat a condition called endometriosis, a condition affecting some 6 to 8 million American women which can cause abnormal bleeding and impair fertility.

Another common but rather different use of carbon-dioxide lasers is to perform delicate microsurgery on the larynx without having to open up the throat. Passing the CO_2 beam down the throat in a small tube lets the surgeon remove small cancers or wart-like growths from the vocal cords without damaging them. Some specialists believe there's no effective substitute for the laser for treating some conditions.

We should emphasize that all surgeons are not about to throw away their scalpels in favor of lasers. The scalpel remains a powerful all-purpose tool, useful in a wide range of surgery, including many types where lasers are not useful. However, the carbon-dioxide laser—and sometimes other lasers—can do a better job than scalpels in some otherwise difficult delicate operations.

Laser Eye Treatment

The first big successes of medical lasers were in eye treatment, and the laser remains an especially valuable tool for ophthalmologists. The simplest laser eye treatment to explain is a rare condition called a detached retina.

The retina is the light-sensitive layer at the back of the eye. Sometimes the retina can become torn, and detached from the back of the eyeball. The damage can spread, and without treatment the entire retina can come loose from the back of the eye, causing blindness. The detachment can be stopped by focusing a laser pulse onto the retina, so it causes a burn which forms scar tissue. Although this damages a small area of the retina, the scar tissue "welds" the retina down to the back of the eyeball, so it cannot break free. The laser treatment represented a big advance over earlier techniques, which required risky open-eye surgery.

A much more common laser procedure is treatment of a condition called "diabetic retinopathy" that is common in people who suffer from juvenile-onset diabetes. For reasons not fully understood, networks of abnormal blood vessels spread across the surface of the retina in many diabetics. These blood vessels are very fragile, and they can leak blood into the normally clear liquid of the eye—gradually dimming vision.

Ophthalmologists have found that illuminating the diseased retina with under a watt of continuous-wave visible light—from an argon, krypton, or dye laser—can slow spread of the abnormal blood vessels. They are not sure exactly how the laser works, but it seems to close off the blood vessels before they can do serious damage. The laser treatment is not a sure cure, but careful studies have shown that it can at least forestall blindness, and it has been used to treat more than 20 million diabetic eyes around the world.

Another type of laser eye treatment is more akin to conventional surgery. It uses a short laser pulse to treat a common complication of cataract surgery. Cataracts occur when the natural lens of the eye becomes cloudy and obstructs vision. They usually occur in older people. The standard treatment is to remove the natural lens material, leaving behind the back membrane of the lens, and to implant a plastic lens. In about a third of all cases, the natural membrane itself can become cloudy, again obstructing vision.

A short, intense pulse from a neodymium laser can pass through the lens implant and break the cloudy membrane, as shown in *Figure 11-5*. The laser pulse is focused with a lens having very short focal length, so the laser light comes to a very tight focus exactly at the membrane surface. The ophthalmologist fires a series of laser pulses to puncture the membrane. Each pulse produces a tiny spark that breaks the membrane. Because the beam is focused very tightly, the power density at the implanted plastic lens is too low to damage it. Behind the membrane, the laser light again spreads out over a large area, so it cannot damage the retina.

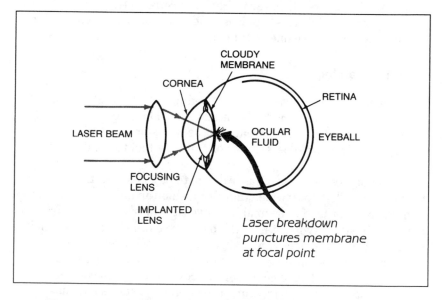

Figure 11-5. A tightly focused laser pulse punctures a cloudy membrane left behind after cataract surgery.

The operation is done in a physician's office, and the membrane opens up almost instantaneously so the patient can see. One leading ophthalmologist says some patients almost jump for joy. Important advantages of the laser treatment include avoiding the time and expense of hospitalization, and the risk of surgically opening the eye to cut the membrane.

Ophthalmologists continue working on new laser treatments for eye disorders. One of the most interesting possibilities is laser "sculpting" of the cornea to correct vision. The cornea is the outermost transparent layer of the eye, shown in *Figure 11-5,* and it accounts for most of the eye's refractive power. Short pulses from ultraviolet excimer lasers can remove material from the cornea to change its refractive power. Researchers hope this might someday correct the vision of people who now wear eyeglasses or contact lenses. However, before you throw away your glasses, you should realize that this is an experimental procedure and not without the risk of complications. Uncertainties include potential long-term complications of removing material from the cornea, which is a thin membrane, and possible hazards from exposure to the short-wavelength ultraviolet laser pulses. There are plenty of reasons for caution—you only have one set of eyes.

Laser Dermatology

Dermatologists, who had long used light to treat skin disorders, were among early medical-laser enthusiasts. The laser has not had as dramatic an impact on skin treatment as it has on eye treatment, but it has given dermatologists an important new tool to treat some conditions.

The biggest success of laser dermatology is in treating dark-red birthmarks called "portwine stains," which often appear on the face or neck. Networks of abnormal blood vessels just under the surface of the skin cause portwine stains; the blood gives the birthmarks their dark color. Because portwine stains are dispersed over the surface of the skin, conventional surgery cannot remove them effectively, and many sufferers had little choice but to cover them with cosmetics.

The standard laser treatment for portwine stains long has been to illuminate them with blue-green light from an argon laser, which is absorbed by the dark-colored blood and heats the blood vessels. The laser beam burns and blisters the skin, but it also causes the blood vessels to close over a number of weeks. As the burns heal, the birthmark bleaches away. While the treatment can be painful, it usually is effective in removing dark birthmarks, although it can cause scarring in some patients, and it cannot be used with young children or people with comparatively pale birthmarks.

A new alternative is the use of a dye laser tuned to 577 nm, the wavelength where hemoglobin in blood has its peak absorption. Because this concentrates absorption in the blood vessels, it avoids the blistering of the skin that accompanies argon-laser treatment. It also is more selective, and thus can be used to treat birthmarks too light for the argon-laser treatment. This, and the avoidance of burning, also allows the treatment of young children, whose birthmarks normally are pale but darken with age.

Laser Cancer Treatment

Lasers are an integral part of a new treatment for cancer called photodynamic therapy. The treatment relies on a dye, called hematopophyrin derivative (HpD), which cancer cells retain much longer than healthy cells.

A patient is given an injection of HpD, then waits for three days while healthy cells flush out the dye, leaving it concentrated in cancer cells. Then the physician illuminates the cancer with red light near 631 nanometers—usually from a gold vapor laser or dye laser—a wavelength where HpD absorbs light strongly but healthy cells absorb light only weakly. The absorbed light makes the dye react chemically with oxygen molecules in the cell, producing a highly reactive form of oxygen that can be lethal to the cancer cell in high enough concentrations.

Photodynamic therapy remains in development, but researchers have reported encouraging results in clinical trials with human patients. It is most effective in treating cancers that can readily be reached with light.

New Medical Treatments

Medical researchers continue to work on new laser techniques for treating disease. Many new concepts use optical fibers to deliver light inside the body without the need for surgery, or use specific wavelengths that have particular interactions with tissue. There isn't room to cover them in any detail, but we can mention a few promising ideas:

- Delivery of light from a dye laser through optical fibers to shatter kidney stones and gallstones into fragments small enough to pass freely from the body. The laser is tuned to a wavelength strongly absorbed by the stones. Prospects for this treatment look very good.

- Delivery of laser light through optical fibers to remove blockages from arteries in the arms, legs, and regions around the heart. Several types of lasers are being studied, including argon, excimer, dye, and neodymium-YAG. Researchers have shown that lasers can remove arterial blockages, but misguided laser light or fibers also can remove arterial walls. Serious concerns remain about damage to the arteries, and some observers worry that the problems may not be solvable for the tangled arteries near the heart.

- Delivery of laser light through a bundle of optical fibers to treat bleeding ulcers in the stomach. The neodymium-YAG laser has been used, but other types might also work.

Cold Laser Treatment

In all the laser treatments we have described so far, the laser beam produces an observable change in something, whether by vaporizing tissue or closing blood vessels. However, there is also a family of "cold" laser treatments that require only a few milliwatts of power from a helium-neon or semiconductor laser. The orthodox medical establishment is skeptical of their effectiveness, but some people have claimed good results. Examples include:

- Laser "acupuncture," in which a low-power laser beam illuminates the acupuncture points originally defined for classical Chinese acupuncture with needles. The effects are said to be similar to needle acupuncture.

- Alleviation of chronic pain by illuminating affected areas with a low-power laser.

- Speeding of wound healing by illumination with a low-power laser beam.

While there has been legitimate controversy over the effects of such cold laser treatments, there also have been some allegations of fraud concerning the use of low-power lasers for "biostimulation." Several years ago, the Federal Trade Commission filed complaints against two Florida practitioners who were making what the FTC considered excessive claims about the benefits of nonsurgical laser "facelifts." The American Society of Plastic and Reconstructive Surgeons made even stronger charges.

PHOTOCHEMISTRY AND ISOTOPE SEPARATION

The energy carried in a light wave or photon can trigger chemical reactions. Light energy can break apart a complex molecule, remove an electron from the outer shell of an atom or molecule, or excite an atom or molecule into a state where it is more ready to engage in chemical reactions.

This raises many interesting possibilities because "photochemical" reactions are very selective. We saw earlier that atoms and molecules have different characteristic absorption wavelengths. This means that light of a specific wavelength can cause one type of molecule to react, but not others present in the same mixture. Thus, selective photochemistry is possible if the light sources are lasers tuned to certain wavelengths.

Suppose, for example, you have a mixture of three chemically similar molecules and you want to excite one of them so it will react with another substance to form a drug. If the desired molecule absorbs light at 400 nm—and the other two absorb at 398 and 402 nm—you need only illuminate the mixture with a 400-nm laser. It will excite the desired molecule so it reacts, but it will not affect the other molecules.

Chemists have developed a formidable arsenal of tools, so they don't need selective laser photochemistry for routine chemical synthesis. However, laser photochemistry is making important strides in the much tougher problem of isotope separation, particularly for the uranium isotopes used in nuclear reactors and the plutonium isotopes used in nuclear weapons.

Isotopes are atoms of the same element that contain different numbers of neutrons in their nucleus but the same number of protons. The number of protons in the nucleus (which equals the number of electrons in a neutral atom) defines the atomic number and hence the identity of an element. For example, all atoms with 92 protons in their nucleus are uranium atoms. However, uranium atoms can have different numbers of neutrons in their nucleus. The two most common uranium isotopes are uranium-238 (with 146 neutrons) and uranium-235 (with 143 neutrons). U-238 is the most common in nature, but it cannot sustain the fission chain reaction needed to drive a nuclear reactor. To sustain a fission reaction, you need the less-common U-235. Conventional fission reactors need 3 or 4% of U-235 to sustain a chain reaction, but the natural U-235 concentration is only about 0.7%. The only way to get those reactors to work is to enrich the U-235 concentration above the natural level.

The problem is that isotopes are chemically identical. Because they have the same number of protons in the nucleus, they have the same electron configurations. The differences in nuclear mass shift the electronic energy levels slightly, but not enough to make their chemical properties different. Conventional isotope enrichment processes have to rely on the small differences in mass between different isotopes.

Laser photochemistry can make the distinction between the two isotopes because lasers can be tuned precisely enough to excite energy levels of one isotope but not the other. *Figure 11-6* shows the concept behind the atomic vapor laser isotope enrichment process for uranium developed at the Lawrence Livermore National Laboratory in California. A dye laser, pumped by a copper-vapor laser, is tuned to a narrow range of wavelengths that excites U-235 atoms but not U-238 in uranium vapor. The excitation process removes an electron from the U-235 atom, leaving it positively charged. That positive ion is attracted to a negatively charged electrode. On the other hand, U-238 atoms retain all their electrons and are not collected by the negative electrode.

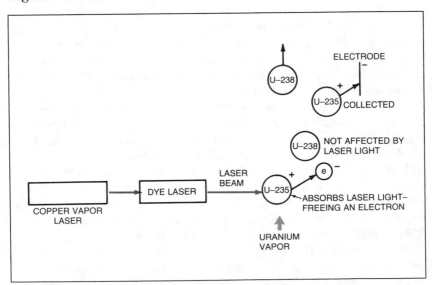

Figure 11-6. Dye laser is tuned to a wavelength that ionizes U-235 atoms but not U-238, allowing concentration of the rarer U-235 isotope.

Figure 11-6 makes it look like the negative electrode only collects U-235, but the process is not that simple—fortunately. We don't really want it to be too easy to produce pure U-235, because the fissionable isotope can be used for nuclear bombs as well as nuclear reactors. In a single pass, the laser process does not produce 100% pure U-235, but it does "enrich" the product in U-235. The process also has other complications, such as the handling of uranium vapor (truly nasty stuff!).

Livermore has developed a similar process for separating isotopes of plutonium, which are produced in nuclear reactors when U-238 atoms absorb neutrons. Plutonium-239, like U-235, is fissionable, and can sustain nuclear chain reactions in reactors or bombs. However, plutonium-240 and -241 isotopes are not fissionable, and tend to "poison" nuclear chain reactions. Reactors also produce the heavier isotopes, and those must be removed to purify the plutonium-239 for further use. In the United States, which has decided not to use plutonium in reactors, that further use is in nuclear weapons.

LASER-DRIVEN NUCLEAR FUSION

Another use of lasers in nuclear energy is in research into nuclear fusion. The fission reactors we use today split the nuclei of heavy atoms such as uranium and plutonium to generate energy. Nuclear fusion, which occurs naturally in the sun and stars, releases energy by combining the nuclei of light elements such as hydrogen and helium.

Fusion has long seemed an attractive energy source. In theory, it should produce less radioactive nuclear garbage than fission reactors, which inevitably leave radioactive fragments of heavy nuclei. Another advantage is that the light nuclei that fuel a fusion reaction are far more abundant on earth than the heavy elements needed to fuel fission reactors.

The bad news has been the difficulty of producing the right conditions for fusion. It takes extremely high temperatures and pressures to force light nuclei together so they can combine and release energy. Those pressures and temperatures are readily available in the core of the sun. We can produce them on earth by exploding a nuclear fission bomb (which serves as the trigger for a hydrogen or fusion bomb). However, to generate power, we want to release energy in a controlled manner, and a bomb is hardly a controlled release of energy.

Researchers trying to control thermonuclear fusion have pursued two approaches. One is magnetic confinement, which tries to keep the hot fusion plasma in a strong magnetic field. The other is inertial confinement, which seeks to compress the fusion fuel into a tiny hot, dense plasma for a short period, so nuclear fusion can occur and release energy. The major approach to inertial confinement fusion is to use intense pulses of laser light.

Figure 11-7 shows the idea of laser fusion. Short, intense pulses of laser light strike a pellet containing hydrogen isotopes that can fuse together to release energy. The laser energy must be distributed uniformly around the surface of the sphere. When the laser energy hits the surface, it produces an implosion, forcing the material inside the tiny sphere inwards and simultaneously heating it to high temperatures. At the peak of the implosion, the nuclei in the pellet fuse together, releasing energy.

The laser power levels needed for fusion are very high—the peak power is measured in trillions of watts (terawatts). The time scale also is very short—laser pulses last around a nanosecond. So far the quantities of fusion energy generated are small, but the process has produced fusion reactions.

Before you get too enthusiastic about laser fusion, a few words of caution are appropriate. Even the optimists consider laser fusion a 21st century technology. It's by no means certain that laser fusion will ever generate power commercially, and if it does, the date probably will be well into the 21st century. In the meantime, the main reason for continuing the program is military research. The implosion of a pellet of fusion fuel involves the same physical processes as the explosion of a hydrogen bomb. Fusion microexplosions let bomb developers study the process on a more convenient scale.

LASER WEAPONS

The simplest way to think of laser weapons is as materials processing on unfriendly objects. The basic processes involved are similar. A laser weapon must do something to the surface of a target, although the specific goal is to make the target inoperative as a weapon rather than to drill a hole. Laser weapons have many potential missions. The one that has gotten the most attention

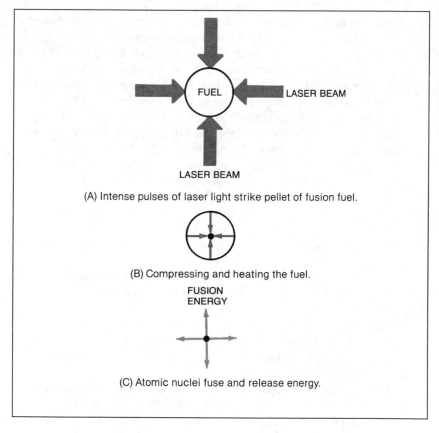

(A) Intense pulses of laser light strike pellet of fusion fuel.

(B) Compressing and heating the fuel.

(C) Atomic nuclei fuse and release energy.

Figure 11-7. Intense laser pulses implode a pellet containing nuclear fuel to generate fusion reactions.

recently is in defense against attack by nuclear missiles. However, high-energy lasers also have been studied for use against satellites, short-distance battlefield missiles, and sensors.

The task of laser weapons is much more complex than materials processing. In materials processing, you try to make objects absorb laser light. In warfare, the other side tries to protect potential targets against laser attack. Military targets are much farther away, making it harder to focus laser energy onto them. They may be coated with materials that deflect laser energy, rather than absorb it. And the targets will be moving in ways that the person (or control system) firing the laser isn't supposed to know. All this adds up to requirements for much higher powers than for materials processing. The most powerful lasers used for materials processing deliver continuous-wave powers of about 15 kilowatts—but 5 to 10 kW is much more common. Precise power levels needed to destroy military targets are classified, but it's clear that average or continuous powers in the megawatts are needed for use against missiles.

Distance is a serious complication for many reasons. The atmosphere is nowhere near as transparent as we think. Air always absorbs a small fraction of the light it transmits. Absorption of only 0.1% of the energy in the beam may seem a small loss, but that represents the addition of significant amounts of heat to the atmosphere for a high-power beam. For a 1-megawatt laser beam, 0.1% is a kilowatt, and that's enough to heat the atmosphere and make it turbulent so it bends the laser beam in unpredictable directions. This turbulence makes it hard both to track the target and to focus the laser beam onto it effectively.

Going into space can overcome atmospheric transmission problems, but it begs the problem of how to get massive hardware into space, and how to provide power for it when it's up there. Moreover, there are hosts of other problems, including the need to identify and track targets and control the entire weapon system effectively. In fact, building a high-energy laser may be one of the lesser of the problems.

In short, there are truly formidable problems to building laser weapons able to destroy nuclear-armed ballistic missiles. The problems are not necessarily insurmountable, but it will take billions of dollars to investigate them, and there is absolutely no assurance that we will be able to overcome them.

We can make the task of building laser weapons appear more manageable if we redefine it to a more limited goal of disabling targets. If we are willing to limit ourselves to knocking out sensors—such as the infrared sensors on guided missiles or the sensitive electronic eyes of spy satellites—much lower laser powers will suffice. Even analysts who doubt the Strategic Defense Initiative will ever produce anything capable of defending against nuclear attack believe that ground-based lasers can pose a serious threat to military satellites, particularly those used for surveillance. The laser rangefinders and target designators now used on the battlefield pose eye hazards for friendly troops, and military planners worry that future anti-sensor lasers might be a more serious threat to human eyes.

What Have We Learned?

- High-power applications change materials which are not unusually sensitive to light. They are less diverse than lower-power applications.

- High-power lasers can concentrate beams to produce extremely high power densities on small areas. They do not apply physical force to the objects they strike.

- Materials working includes cutting, welding, drilling, and marking both metals and nonmetals.

- Lasers can drill holes in diamond, the hardest substance known or in flexible materials like baby-bottle nipples without deforming them.

- Laser machining is a noncontact process.

- Energy transfer from the laser beam depends on wavelength and material absorption.

- Most metals absorb little light at CO_2 laser wavelengths.

- Plasma released from an illuminated object's surface can block the laser beam.

- Lenses with longer focal length have greater depth of focus so they can cut or drill deeper.

- Drilling requires pulsed lasers and depends on peak power.

- Lasers can mark or scribe materials without drilling holes all the way through.

- Lasers cut with a continuous beam. Metal cutting usually depends on oxygen to remove material.

- Welding joins two pieces into a single unit.

- Heat treating transforms the surface of a metal to a harder form.

- Power levels for materials working range up to kilowatts.

- Lasers are widely used in making semiconductor electronics.

- Lasers are widely used in medicine, but fundamental research continues on laser-tissue interactions.

- Tissue reaction to light depends on wavelength.

- The 10-μm CO_2 laser can vaporize tissue and prevent bleeding.

11

- Visible and near-infrared lasers have other medical uses.
- Lasers can treat detached retinas and diabetes-related blindness.
- Lasers can treat complications from cataract surgery, and someday lasers may be able to correct defective vision.
- Laser photodynamic therapy can destroy some cancers.
- Laser light can trigger chemical reactions.
- Laser photochemistry can separate isotopes of uranium and plutonium for nuclear reactors and bombs.
- In laser fusion, intense laser pulses heat and compress a fusion-fuel target. Fusion has been demonstrated, but commercial reactors are a long ways away.
- High-power lasers seem attractive as military weapons, but the technology is very difficult. Antisatellite and antisensor laser weapons are much easier to build than antimissile lasers.

Quiz for Chapter 11

1. Which of the following is not an attraction of high-power lasers?
   a. Ability to generate extremely high powers in very short pulses
   b. Amenable to robotic control
   c. Do not apply physical force to objects
   d. Are extremely efficient
   e. Can focus light energy tightly onto a small spot to generate very high powers

2. Which of the following is most important for laser drilling of baby-bottle nipples?
   a. Ability to generate extremely high powers in very short pulses
   b. Amenable to robotic control
   c. Do not apply physical force to objects
   d. Are extremely efficient
   e. Can focus light energy tightly onto a small spot to generate very high powers

3. A 1000-watt continuous-wave carbon-dioxide laser illuminates a titanium sheet for one second. How much energy does the titanium absorb (assuming that absorption does not change with heating)?
   a. 100 watts
   b. 80 joules
   c. 120 joules
   d. 800 joules
   e. None of the above

4. Which of the following are desirable for drilling holes in a thick (2-cm) slab of material?
   a. Short laser pulses
   b. Short focal-length lens
   c. Long focal-length lens
   d. A and B
   e. A and C

5. Which of the following are not important in producing a good laser weld?
   a. Short laser pulses
   b. Penetration of the laser beam into the joint
   c. Close fitting of the pieces being welded
   d. Compositions of the pieces being welded
   e. Scanning the laser beam along the joint at the right speed

6. Which of the following lasers produces a beam that is absorbed most efficiently by tissue?
   a. Argon-ion
   b. Neodymium-YAG
   c. Ruby
   d. Carbon-dioxide
   e. Semiconductor diode (GaAlAs)

7. What is the most important medical use of the carbon-dioxide laser?
   a. Eye examinations
   b. Treatment of bleeding ulcers
   c. Removal of thin layers from blood-rich tissue
   d. Cutting bone
   e. Removal of portwine stain birthmarks

8. Lasers are useful in isotope separation because they do which of the following?
   a. Provide extremely narrow linewidth light to excite one isotope but not another
   b. Can deliver very short pulses
   c. Concentrate power onto one atom but not adjacent atoms
   d. Heat uranium metal until it is vaporized
   e. Produce thermonuclear fusion

9. What must laser beams do to cause fusion reactions?
   a. Provide extremely narrow linewidth light to excite one isotope but not another
   b. Compress fusion fuel to very high densities while heating it to high temperatures
   c. Push atomic nuclei together
   d. Heat uranium isotopes
   e. All the above

10. What types of targets are likely to be threatened by laser weapons first?
   a. Nuclear-armed ballistic missiles
   b. Extraterrestrial space ships carrying hostile aliens
   c. Sensors and spy satellites
   d. Space stations
   e. Helicopters on the battlefield

Laser Safety

There are two main potential hazards associated with lasers: the beam and the power supply. Even low-power laser beams can pose some dangers to the eye, while high-power laser beams can burn the skin. The power supply, like any source of high voltage—or even wall current—can kill.

We have seen that a laser beam is made up of parallel light rays. The concentration of parallel light rays in a small beam makes laser powers as low as the milliwatt range a potential hazard to your eyes. The reason is that the eye focuses those rays, concentrating their optical power in a small enough area that they can burn the retina. The hazard of a few-milliwatt visible laser beam is comparable to that from looking directly at the sun. An accidental glance past the sun may dazzle your eyes, but staring directly at the sun can cause permanent eye damage. The same is true for lasers. The safest rule is never to look directly into any laser beam, even if you think it has very low power.

Higher laser powers pose more serious eye hazards. As continuous-wave power levels rise, it takes less time for the light to do lasting damage to the eye. If the power is concentrated in short high-power pulses, even a single shot can do lasting damage. One laser scientist has described how a direct hit from a stray pulse made blood erupt inside his eye. Such an event is not likely to blind you totally, but it can leave a blind spot that impairs your vision.

Both visible and invisible laser light poses hazards to your eyes. Invisible near-infrared and near-ultraviolet light can enter your eye, and near-infrared light (in particular) can reach your retina without you sensing it. Hazards become less at wavelengths longer than about 1.55 micrometers, because the light cannot reach your retina, but you nonetheless should exercise care when working with any laser. Some near-infrared lasers can be very deceptive because they seem to be emitting extremely weak red beams, but are actually emitting significant powers at wavelengths which your eyes sense very inefficiently.

Laser safety goggles are made to block the wavelengths emitted by specific lasers and transmit other light. For example, goggles made to block the 1.06-micrometer neodymium laser wavelength transmit visible light; those made to block the argon-laser lines at 511 and 488 nm transmit red light. Good goggles should block light from reaching your eye via any side paths, such as accidental reflections from the side. Remember that it is vital to use goggles made for the type of laser you're using. Goggles made for the red helium-neon wavelength do not block light from an argon laser.

In more than a dozen years of writing about the laser industry, I have never known anyone to be killed by a laser beam. However, several people have been killed when they accidentally touched high-voltage sources within a laser. It is much too easy to take electricity for granted. Even when the power is off, components such as capacitors in high-voltage power supplies can retain a lethal dose of electricity for a long time. Watch out. (This homily is dedicated to the memory of Gordon Brill, a laser-industry veteran who was killed while servicing a copper-vapor laser.)

If you are going to work regularly with lasers, do yourself a favor and become familiar with how to use them safely. Information is available from the federal agency charged with regulating laser safety: The Center for Devices and Radiological Health, 5600 Fishers Lane, Rockville, MD 20857. The Lawrence Livermore National Laboratory has prepared a helpful booklet, *A Guide to Eyewear for Protection from Laser Light* for people using lasers. You can request a copy from Corrine Burgin, Scientific Editor, L-399, Lawrence Livermore National Laboratory, P.O. Box 808, Livermore, CA 94550.

Constants, Conversions, and Symbols

CONSTANTS

Speed of light in vacuum (c)	2.99792458×10^8 meters/sec
Planck's constant (h)	$6.6260755 \times 10^{-34}$ joule-second
Planck's constant (h)	$4.1356692 \times 10^{-15}$ electronvolt-second
Boltzmann constant (k)	1.380658×10^{-23} J/K (joule/Kelvin)
Boltzmann constant (k)	8.617385×10^{-5} eV/K
Rydberg constant (R)	1.0973731534×10^7 per meter
Rydberg constant (R)	13.6056981 eV

CONVERSIONS

Electronvolts to joules	1 eV $= 1.60217733 \times 10^{-19}$ joule
A 1-eV photon has frequency	$2.41798836 \times 10^{14}$ Hertz
A 1-eV photon has wavelength	1.23984 micrometers
Mass of electron	$9.1093897 \times 10^{-31}$ kilogram
Mass of proton	$1.6726231 \times 10^{-27}$ kilogram

SYMBOLS

λ (Greek lambda)	wavelength
n	refractive index
ν (Greek nu)	frequency
ω (Greek omega)	angular frequency, $= 2\pi\nu$
c	speed of light
h	Planck's constant

Glossary

Acousto-optic: Interaction between an acoustic wave and a light wave, used in beam deflectors, modulators, and Q switches

Active medium: The light emitting material in a laser, sometimes used to identify a specific atom in a crystal.

Alexandrite: A synthetic crystal doped with chromium to form a solid-state laser that emits near-infrared light.

Angstrom (Å): A unit of length, 0.1 nanometer or 10^{-10} meter. Its most common use is to measure wavelength in the visible spectrum. Abbreviated Å

Arc lamp: A high-intensity lamp in which an electric discharge continuously produces light.

Attenuation: Reduction of light intensity, or loss. It can be measured in optical density or decibels as well as percentage or fraction of light lost.

Attenuator: An optical element which transmits only a given fraction of incident light.

Average power: The average level of power in a series of pulses. It equals pulse energy times the number of pulses divided by the time interval.

Beamsplitter: A device which divides incident light into two separate beams, one reflected and one transmitted.

Birefringent: Has a refractive index that differs for light of different polarization.

Bistable optics: Optical devices with two stable transmission states.

Brewster's angle: The angle at which a surface does not reflect light of one linear polarization.

G

Chemical laser: A laser which is excited by a chemical reaction. The most common type produces hydrogen fluoride or deuterium fluoride.

Circular polarization: Light polarized so the polarization vector rotates periodically without changing in absolute magnitude, describing a circle. Circularly polarized light can be considered as two equal-intensity linearly polarized beams, one 90° in phase ahead of the other.

Coating: Material applied in one or more layers to the surface of an optical element to change the way it reflects or transmits light.

Coherence: Alignment of the phase and wavelength of light waves with respect to each other

Collimate: The act of making light rays parallel.

Concave: Curving inwards, so the central parts are deeper than the outside, like the inside of a bowl.

Continuous wave: Emitting a steady beam.

Convex: Curving like the outside of a ball so the outer parts are lower than the center.

Cycles per second: The number of oscillations a wave makes in one second; the frequency. 1 cycle per second equals 1 hertz.

Decibel: A logarithmic comparison of power levels, abbreviated dB, and defined as the value: $10 \log_{10}(P_2/P_1)$—i.e., 10 times the base-10 logarithm of the ratio of the two power levels.

Detector: A device which generates an electric signal when illuminated by light.

Difference frequency: An output wave with a frequency equal to the difference in the frequencies of two input waves.

Diffraction: The scattering or spreading of light waves when they pass an edge.

Diode: An electronic device which preferentially conducts current in one direction but not in the other. Semiconductor diodes contain a p-n junction between regions of different doping which lets current to flow in one direction but not the other. Diodes can be emit light (e.g., laser diodes) or detect it.

Diode laser: A semiconductor diode laser in which the injection of current carriers produces light by recombination of holes and electrons at the diode junction between p and n doped regions.

Divergence: The angular spreading of a laser beam with distance.

Electro-optic: The interaction of light and electric fields, typically changing the light wave. Used in some modulators, Q switches, and beam deflectors.

Electromagnetic radiation: Waves made up of oscillating electrical and magnetic fields perpendicular to one another, travelling at the speed of light. Waves can also be viewed as being photons, or quanta of energy. Electromagnetic radiation includes radio waves, microwaves, infrared waves, visible light, ultraviolet radiation, X rays, and gamma rays.

Electronic transition: A change in the energy level of an electron.

Excimer laser: A pulsed ultraviolet laser in which the active medium is a short-lived molecule containing a rare gas such as xenon and a halogen such as chlorine.

Far-infrared laser: One of a family of gas lasers emitting light at the far infrared wavelengths of 30 to 1000 micrometers.

Free-electron laser: A laser in which stimulated emission comes from electrons passing through a magnetic field that varies in space.

Frequency: For light waves, the number of wave peaks per second passing a point. Measured in hertz (cycles per second).

Fused silica: Synthetic silica (SiO_2) formed from highly purified materials.

Gallium aluminum arsenide: A semiconductor compound used in LEDs, diode lasers, and certain detectors. Chemically $Ga_{1-x}Al_xAs$, where x is a number less than one.

Gallium arsenide: A semiconductor compound used in LEDs, laser diodes, detectors, and electronic components, chemically GaAs.

Glass: An amorphous solid, typically made mostly of silica (SiO_2) unless otherwise identified. Silica glasses are transparent to visible light.

Harmonic generation: The multiplication of the frequency of a light wave.

Hertz: The unit of measurement for frequency. 1 hertz (Hz) equals 1 cycle per second.

Heterojunction: A boundary between semiconductors that differ in composition, such as GaAs and GaAlAs.

Index of refraction: The ratio of the speed of light in vacuum to the speed of light in a material, a crucial measure of a material's optical characteristics. Usually abbreviated n.

Indium gallium arsenide phosphide: A semiconductor material used in lasers, LEDs, and detectors. The band gap—and hence the wavelength emitted by light sources and detected by detectors—depends on the mixture of the four elements. Abbreviated InGaAsP.

Infrared: Invisible wavelengths longer than 700 nm and shorter than about one millimeter. Wavelengths from about 700 nm to 2000 nm are called the near infrared.

InGaAsP: Indium gallium arsenide phosphide, a semiconductor compound used in light sources and detectors. The band gap depends on composition. It can be written $In_{1-x}Ga_xAs_{1-y}P_y$, where x and y are numbers less than one.

Injection laser: Another name for semiconductor laser. The name comes from the fact that the injection of current carriers leads to the production of light.

Integrated optics: Optical elements analogous to integrated electronic circuits, with multiple devices on one substrate.

Intensity: Power per unit solid angle.

Interference: The addition of the amplitudes of light waves. In destructive interference the waves cancel; in constructive interference they combine to make more intense light.

Irradiance: Power per unit area.

Junction laser: A semiconductor diode laser.

Laser: Acronym for Light Amplification by Stimulated Emission of Radiation. One of the wide range of devices that generate light by that principle. Laser light is directional, covers a narrow range of wavelengths, and is more coherent than ordinary light.

LED: Acronym for Light-Emitting Diode. A semiconductor diode which emits incoherent light by spontaneous emission.

Light: Strictly speaking light is electromagnetic radiation visible to the human eye. Commonly the term is also applied to electromagnetic radiation close to the visible spectrum that acts similarly, including the near infrared and near ultraviolet.

Light-emitting diode: A semiconductor diode which produces incoherent light by spontaneous emission. Usually called an LED.

Longitudinal modes: Oscillation modes of a laser along the length of its cavity, so twice the length of the cavity equals an integral number of wavelengths. Distinct from transverse modes, which are across the width of the cavity.

Maser: Microwave analog of a laser. an acronym for Microwave Amplification by Stimulated Emission of Radiation.

Micrometer: One millionth of a meter, abbreviated μm. (μ is the Greek letter mu).

Mode: A manner of oscillation in a laser. Modes can be longitudinal or transverse.

Monochromatic: Containing only a single wavelength or frequency.

Multimode: Containing multiple modes of light. Typically refers to lasers that operate in two or more transverse modes.

n region: Part of a semiconductor doped so it has an excess of electrons as current carriers.

Nanometer: A unit of length equal to 10^{-9} meter. Its most common use is to measure visible wavelengths.

Nanosecond: A billionth of a second, 10^{-9} second.

Near infrared: The part of the infrared spectrum nearest the visible spectrum, typically 700 to 1500 or 2000 nm, but not rigidly defined.

Normal (angle): Perpendicular to a surface.

G

Oscillator: A laser cavity with mirrors so stimulated emission can oscillate within it. To some purists, only an oscillator can be a laser.

p region: Part of a semiconductor doped with electron acceptors in which "holes" (vacancies in the valence electron level) are the dominant current carriers.

Peak power: Highest instantaneous power level in a pulse.

Phase: Position of a wave in its oscillation cycle.

Photodetector: A light detector.

Photodiode: Usually, a semiconductor diode which produces an electrical signal proportional to light falling upon it. There also are vacuum photodiodes which can detect light.

Photometer: An instrument for measuring the amount of light visible to the human eye.

Photons: Quanta of electromagnetic radiation. Light can be viewed as either a wave or a series of photons.

Polarization: Alignment of the electric and magnetic fields that make up an electromagnetic wave. Normally this refers to the electric field. If light waves all have a particular polarization pattern, they are called polarized.

Polarization vector: A vector indicating the direction of the electric field in an electromagnetic wave.

Polarizer: A device that transmits light of only one polarization.

Population Inversion: The condition when more atoms are in an upper energy level than in a lower one. A population inversion is needed for laser action.

Pumping: The method by which a laser gets the energy to produce a population inversion.

Q switch: A device which changes the Q (quality factor) of a laser cavity to produce a short, powerful pulse.

Quartz: A natural crystalline form of silica.

Quaternary: A compound made of four elements, e.g., InGaAsP.

Radiant flux: Instantaneous power level in watts.

Radiometer: An instrument to measure power (watts) in electromagnetic radiation. Distinct from a photometer.

Rays: Straight lines which represent the path taken by light.

Recombination: Capture of a conduction-band electron by a "hole" in a semiconductor, causing the conduction band to drop to a vacancy in the valence band.

Refraction: The bending of light as it passes between materials of different refractive index.

Refractive index: The ratio of the speed of light in vacuum to the speed of light in a material. A crucial measure of a material's optical characteristics. Abbreviated n.

Resonator: A region with mirrors on the ends and a laser medium in the middle. Stimulated emission from the laser medium resonates between the mirrors, one of which lets some light emerge as a laser beam.

Semiconductor diode laser: A laser in which recombination of current carriers at a p-n junction generates stimulated emission.

Silica: Silicon dioxide, SiO_2, the major constituent of ordinary glass.

Silica glass: Glass in which the main constituent is silica.

Single-frequency laser: A laser which emits only a very narrow range of wavelengths, nominally a single frequency but actually a very narrow range small enough to be considered a single frequency.

Single-mode: Containing only a single mode. Beware of ambiguities because of the difference between transverse and longitudinal modes. A laser operating in a single transverse mode typically does not operate in a single longitudinal mode.

Speckle: Coherent noise produced by laser light. It gives a mottled appearance to holograms viewed in laser light.

Spontaneous emission: Emission of a photon without outside stimulation when an atom or molecule drops from a high energy state to a lower one.

Stimulated emission: Emission of a photon that is stimulated by another photon of the same energy; the process that makes laser light.

Ternary: Compound made of three elements, e.g., GaAlAs.

III-V semiconductor: A semiconductor compound made of one (or more) elements from the IIIA column of the periodic table (Al, Ga, and In) and one (or more) elements from the VA column (N, P, As, or Sb). Used in LEDs, diode lasers, and detectors.

Threshold: The excitation level at which laser emission starts.

Threshold current: The minimum current needed to sustain laser action, usually in a semiconductor laser.

Time response: The time it takes to react to a change in signal level.

Total internal reflection: Total reflection of light back into a material when it strikes the interface with a material having lower refractive index at a glancing angle.

Transition: Shift between energy levels.

Transverse modes: Modes across the width of laser. Distinct from longitudinal modes, which are along the length.

Tunable: Adjustable in wavelength.

Ultraviolet: Part of the electromagnetic spectrum at wavelengths shorter than 400 nm, to about 10 nm, invisible to the human eye.

Vibronic: An electronic transition accompanied by a change in vibrational energy level.

Visible light: Electromagnetic radiation visible to the human eye, at wavelengths of 400 to 700 nm.

Waveguide: A structure which guides electromagnetic waves along its length. An optical fiber is an optical waveguide.

Wavelength: The distance an electromagnetic wave travels during one cycle of oscillation. Wavelengths of light are measured in nanometers (10^{-9} meter) or micrometers (10^{-6} meter).

YAG: Yttrium aluminum garnet, a crystalline host for neodymium lasers.

Index

I

Answers to Quizzes

Chapter 1

1. c
2. b
3. a
4. e
5. b
6. d
7. a
8. b
9. d
10. b

Chapter 2

1. b
2. c
3. a
4. d
5. c
6. d
7. a
8. b
9. d
10. c
11. d
12. c

Chapter 3

1. b
2. e
3. b
4. c
5. d
6. e
7. a

8. d
9. c
10. a

Chapter 4

1. b
2. e
3. a
4. e
5. c
6. e
7. b
8. d
9. b
10. a

Chapter 5

1. a
2. d
3. d
4. b
5. c
6. e
7. c
8. a
9. c
10. e

Chapter 6

1. c
2. a
3. b
4. d
5. d

6. a
7. e
8. c
9. e
10. d

Chapter 7

1. e
2. b
3. a
4. d
5. b
6. e
7. c
8. c
9. a
10. d

Chapter 8

1. c
2. a
3. c
4. b.
5. c
6. b
7. d
8. b
9. a
10. e

Chapter 9

1. d
2. d

3. a
4. b
5. b
6. a
7. d
8. c
9. d
10. d

Chapter 10

1. e
2. c
3. a
4. b
5. e
6. d
7. b
8. c
9. c
10. d

Chapter 11

1. d
2. c
3. b
4. e
5. a
6. d
7. c
8. a
9. b
10. c

21:80